Chemtrails, HAARP, and the Full Spectrum Dominance of Planet Earth

10 9 8 7 6

Feral House
1240 W. Sims Way
Suite 124
Port Townsend WA 98368

www.feralhouse.com

Design by Jacob Covey

Chemtrails, HAARP, and the Full Spectrum Dominance of Planet Earth

ELANA FREELAND

FERAL HOUSE

ACKNOWLEDGEMENTS

Thank you to Diana Thatcher for loaning me books from her extensive library; Mary Rooney for her ongoing research from the UK; Felicia Trujillo, ND, for her intrepid EM research in the face of harassment; Rose at The CON Trail for providing an Internet forum for chemtrails networkers, by which I learned of the extraordinary passion of Australian/New Zealand anti-chemtrails activists; Wayne Hall in Greece for his dogged commitment to anti-chemtrails activism; Talitha Talia for her acute weather-watching and health tips; Carolyne Hart for taking a struggling writer out to dinner now and then; and a special thanks to publisher Adam Parfrey for suggesting that I write the book, and to independent scientist Clifford E. Carnicom for his patience and openness in discussing all things aerosol.

Table of Contents

INTRODUCTION

Navigating the National Security State

We are as gods and might as well get good at it.

— Stewart Brand, *Whole Earth Catalog*, 1969

Rough weather is rolling into Houston from the south today, proving that those alleged government-controlled weather machines aren't affected by the [2013 government] shutdown. Already pictures of menacing-looking clouds are making the rounds on social media.

— Craig Hlavaty, "Strange cloud formation grabs Houston's attention as storms roll in," *Houston Chronicle*, October 1, 2013

I can't imagine stratospheric aerosols ever being deployed . . . You can't test it unless you basically do full-scale deployments . . . I just can't see the world standing for that. You would have to notify everybody that might be affected—informed consent over the entire planet—and you'd have to do an Environmental Impact Statement and I can't imagine everybody in the world agreeing to those changes . . .

— Alan Robock, Distinguished Professor of Climatology, Rutgers University, 2012

Planet Earth.

The first Earth Day on April 20, 1970 seems eons ago—the brainchild of Senator Gaylord Nelson (D-WI) and the beginning of an environmental decade loaded with green legislation and a slew of new federal agencies supposedly created to implement the Occupational Safety and Health Administration: the Clean Air Act, National Environment Policy Act, Clean Water Act, Noise Control Act, Endangered Species Act, the Superfund, and Alaska National Interest Lands Act.

Almost half a century later, Earth Day rolls on but those once-hopeful agencies seem to be serving corporate agendas that destroy, not protect, the

Earth. By 2004, BBC News was quoting from "The Earth's Threatened Life-Support System: A Global Wake-Up Call" regarding uncharted waters of an Anthropocene epoch, "the geologic epoch in which humans are a significant and sometimes dominating environmental force"

> . . . sailing into planetary terra incognita . . . unsure of just how serious our interference with Earth system dynamics will prove to be . . .[1]

all of which sounds more like the High-frequency Active Auroral Research Project (HAARP) and chemtrails technology than carbons and greenhouse gases. Then Stanford University population biologist Paul Ehrlich, author of the 1968 book *The Population Bomb,* was resurrected to reiterate the old saw of blaming overpopulation for imminent global collapse in the prestigious British science journal *Proceedings of the Royal Society.*[2] Carbons and overpopulation, but not one mention of *stratospheric aerosol geoengineering (SAG), solar radiation management (SRM),* or the geophysical impact of chemtrails and ionospheric heaters like HAARP.

The truth is that we are entering a technological Space Age very different from the one envisioned by President John F. Kennedy. If the military mindset has its way, all of Planet Earth will be militarized under the policy of *full spectrum dominance.* One-eyed mainstream media are on board to task the public with pollution and "global warming." Like a sonar echo, the military bemoans "battlefield attrition" and blames what it calls the "CNN syndrome" (the public's desire to be informed) for "forcing" the U.S. Army "to look/act like SOCOM [Special Operations Command]," meaning steeped in secrecy, deceit, and plausible deniability. The "CNN syndrome" public simply cannot handle the necessity of weaponizing everything under the Sun, from "ubiquitous multiphysics, hyperspectral sensors, precision strikes, volumetric weaponry, swarms and hardened munitions" to "non-explosive warfare" like psywar, biowar, IT/net war, "anti-operability war," beam weaponry, RF, spoofing/camo, robotic warfare "in the large"/better than human AI/"Cyber life," and "alternative power projection approaches (e.g., the deep water depth/death sphere)."[3]

Ever so quietly, the science of waging war has been transformed by the Revolution in Military Affairs (RMA). The "battlespace" is everywhere, foreign and domestic, and the IT/Bio/'Bot weapons of choice require an ionized (electrified) atmosphere. Wars over natural resources are being replaced by wars

---

1 Alex Kirby, "Earth 'entering uncharted waters'," *BBC News,* 20 January 2004. Authors of the article in *Global Change and the Earth System: A Planet Under Pressure* (Springer, 2005) were Paul Crutzen, Ph.D., 1995 Nobel prize for chemistry; Bert Bolin, Ph.D., founding chair of the Intergovernmental Panel on Climate Change; Margot Wallstrom, European Environment Commissioner; Dr. Will Steffen, director of the International Geosphere-Biosphere Programme.

2 Stephen Leahy, "Experts Fear Collapse of Global Civilization." Inter Press Service, January 11, 2013.

3 Quoted from the 113-page NASA Langley Research Center PowerPoint document *The Future Is Now! Future Strategic Issues/Future Warfare [Circa 2025],* written by Bill Stryker of DIA/Futures, July 2001.

over "societal disruptions" as the line between civilian and soldier is erased. *Directed energy weapons (DEWs)* promise "warfare on the cheap" (steerable avoidance of collateral damage), survivability ("Can see everything, anything you can see you can kill"), and effectiveness ("Lethality of precision and volumetric weaponry").[4]

HAARP technology with its ionized chemtrails constitutes a global-scale DEW system.

While mainstream newscasters wring their hands over "global warming" and "climate change," ionospheric heaters torque the chemtrails and unnaturally heat the planet in endless military "experiments." The President promises "new climate proposals" as TransCanada Corp's $5.3 billion Keystone XL pipeline is pushed through Congress to deliver tar-sands oil that promises yet more greenhouse gas emissions.[5] Meanwhile, stealthy aflatoxins, Ebola and Lassa, binary agents, and genomic-targeting pathogens ride nanoparticulates into our lungs and bloodstreams in a "long-term fingerprintless campaign."[6]

The "Worldwide IT Revolution" is synonymous with a C4 (command, control, communications, computers) weapons system. Thanks to "superb worldwide sensor suites and precision strike capabilities," it's "tele-everything." Communications, computing, and sensors are silicon, molecular, quantum, bio, and optical, soon to move humans into AIs and "Automatics / Robotics 'in the large'," immersive multisensory VR (virtual reality "holodecks"), and multiphysics hyperspectral sensors in land, sea, air, and space. By *circa* 2025, robotics will be the norm, with binary bios in the agriculture and food distribution systems, with nanosensors, munitions, and swarm/horde weapons keeping a brave new virtual world peace tacked to a wireless Smart Grid "Cloud."

Such is our transitional condition in this 67th year of the American National Security State.

1

Secrecy always accompanies cutting-edge technologies yoked to war.

In 2010, 76.7 million documents were classified, compared with 8.6 million in 2001[7]. While the military plays its part, CEOs and boards of directors behind the closed doors of vast transnational corporations connive for more and more power and profits at the expense of human-scale society, and government institutions mandated to serve the people are now more alien than Big Brother in the dystopic classic *1984*.

---

4 Ibid.

5 Lisa Lerer, "Obama Tells Keystone Foes He Will Unveil Climate Measures." *Bloomberg*, June 14, 2013.

6 *The Future Is Now!* "Some Interesting 'Then Year' BW [biological warfare] Possibilities," 2001.

7 Andy Greenberg, *This Machine Kills Secrets: How WikiLeakers, Cypherpunks, and Hacktivists Aim to Free the World's Information*. Dutton, 2012.

The aerosol program going on in the sky over our heads remains secret because for the first time in human history, we are being forced to live in a chemicalized atmosphere more like a plasma "battery" than the sky our forebears enjoyed. Global powerbrokers monitor and map our every movement while subjecting all of biological life to ionized aerosols loaded with experimental biological and chemical agents. As *Aviation Week*'s senior international defense editor Bill Sweetman put it in his 1993 book *Aurora: The Pentagon's Secret Hypersonic Spyplane*: "Cover and covert are the principal words in the secret world's vocabulary . . . Lifting the cover on intelligence systems reduces their value."

The wartime secrecy of World War II and the Manhattan Project was continued as the Cold War with the passage of the National Security Act of 1947 and conversion of the wartime Office of Strategic Services (OSS) into the Central Intelligence Agency (CIA). In 1951, the Inventions Secrecy Act was added to conceal technological inventions and shift the patent system from open source to control by the new National Security State. To this day, every electromagnetic technology that can be weaponized ends up under a shroud of "national security."

The American people did not foresee the impact that blanket secrecy would have on their peacetime representative government. They thought they understood the whys and wherefores of Cold War containment of military and intelligence secrets, including compartmentalization and need-to-know clearances. Little did they realize that cabals, dynastic corporations, and crime syndicates—experts in secrecy, lies and deception, mayhem and murder—would gather like vultures to double-cross elected officials and split society between those burying their heads in the flag and those struggling to penetrate how 24/7 canned "news," disjointed factoids, and social media peppered with memes were undermining the social order. Maintain a lie or secret for one generation and the next generation becomes easy pickings, particularly if the lie is echoed in school textbooks, TV cartoons, the History and Discovery Channels. Confuse one generation and the next will happily ingest the media gruel prepared for them.

An army of sociology and psychology Ph.D.s like Austrian-American public relations consultant Edward Louis Bernays (1891–1995) and German-American psychologist Kurt Zadek Lewin (1890–1947) has designed a system of information overload and cognitive dissonance guaranteed to desensitize Americans. Deluge people with bad "news" and uncertainty in tiny contextless "news bites" and trigger the emotional overwhelm that will drive them to seek one distraction and convenience commodity after another, thus enriching corporations while making "we the people" accept more and more secret programs for our own good.

A real conspiracy is underway, composed of Air Force personnel, weather forecasters, civil air traffic controllers, pilots, media, environmental agencies and associations, senators, representatives, physicians, and scientists in denial of service and breach of duty to inform and protect the public. The artificially mediated environment spoon-fed to us by "experts" has crushed much of our confidence in our ability to perceive what is really going on right before our eyes. Many technical secrets of this conspiracy are in public sources, but we do not know how to interpret them, given that our authority figures and "experts," in one way or another, have been bought off and silenced.

The subterfuge surrounding "geoengineering" is thick as molasses—from Al Gore's *An Inconvenient Truth* (2006) to fake "global warming" NGOs (nongovernment organizations) rising up like chimera to blame the public, broker deals with geoscientists, and manipulate scientific data. The California Air Resources Board claimed *it had no air quality records for 2002–2011*, and when data was finally released to Environmental Voices (www.environmentalvoices.org), it had been reworked. U.S. science adviser John Holdren insists that releasing particulates of barium, magnesium, aluminum, nanofibers, mold and bacillus blood spores is a great idea.[8] After all, what are chemtrails but a global extension of biological and chemical warfare (BW/CW) field experiments that have been going on for decades?

We find ourselves in an upside-down world in which government and industry profit from creating disasters and running biological "trial experiments" on the people whose tax dollars feed their children—from the Nuclear Regulatory Commission's Atomic Safety and Licensing Board's approval of Southern California Edison's plan to continue operating the collapsing San Onofre nuclear power plant to "see what happens,"[9] to the New York Police Department ("working with Long Island's Brookhaven National Laboratory") releasing perfluorocarbon gases into the subway system during morning rush hour "to study how chemical weapons could be dispersed through the air."[10]

Under such a regime of secrets, lies, and plausible deniability, our greatest defense must begin with learning how to read between the lines of mainstream media smoke and mirrors so as to discern how public inquiry is being confused and marginalized by assaults like that of the overused, hackneyed term "conspiracy theory." Lay researchers with no tenure to protect nor fact to hide may now be our only hope. As I said, much can be found in open source, if we but have the *context* with which to interpret it. Waiting for NASA or military-supported university labs and their phalanxes of "experts" to tell us the truth about what is going on in our skies is folly. Nor should we depend upon honest whistleblowers or canaries in the coal mine to brave charges of treason and even

---

8 Susanne Posel, "How Gov Uses Geoengineering Experiments Without Telling You." *Occupy Corporatism*, June 17, 2013.

9 Fortunately, this attempt failed and San Onofre is being shut down, which Edison expects customers to pay for: Marc Lifsher, "Edison tells customers they should pay for San Onofre shutdown." *Los Angeles Times*, August 12, 2013.

10 "NYPD Releases Harmless Gases in Subway for Chemical Weapon Test." *NBCNewYork*, July 9, 2013.

death. It is time to concentrate on working together to discern true from false, information from disinformation—a truly democratic process.

2

In the late 1990s, when HAARP was coming online and jets were laying "chembombs" that made people below violently ill, the Internet was alive with guessing games about the strange behavior of "persistent contrails" lingering and dispersing into what NASA Langley Research Center scientists had already coined as *cirrus contrailus*.[11] At the same time, a meme utilized by the U.S. Air Force Academy[12] flamed over the global Internet: *chemtrails*.

According to *Chemtrails Confirmed* (2004) by William "Will" Thomas, a former pilot and ocean navigator,[13] the earliest chemtrail observers were World War II and Korean War veterans who had retained their habit of keeping a sharp eye on the sky. Immediately, they recognized that the skies over their California homes were being "gridded," and *not* by normal contrails. When they contacted Military Operation Control at nearby Travis Air Force Base and Fresno Yosemite International Airport, all they got for their trouble was plausible deniability: no flight plans equals no jets overhead.

Researchers slowly began to realize that some of the "friendlies" they were encountering on the Internet were spreading the bad seed of *disinformation*[14] and that they were on their own, given that there were no topical books in the library—at least until 1995 when the book *Angels Don't Play This HAARP* by Dr. Nick Begich and Jeane Manning came out. "Insiders," "anonymous sources," and whistleblowers began slipping out of the shadows. Were they legitimate, or disinformation agents? Kevin Barrett, Ph.D., offers insight into the problem:

> The main way that covert operators keep secrets is not by keeping them, but by revealing them—through a discredited source. For example, young George W. Bush's cocaine arrest, which got him thrown out of the National Guard, was revealed (in an operation orchestrated by Karl Rove) to a journalist named Jim Hatfield. Rove, Bush's minister, and others confirmed the story to Hatfield.

---

11 Patrick Minnis, J. Kirk Ayers, Steven P. Weaver, "Surface-Based Observations of Contrail Occurrence Frequency Over the U.S., April 1993-April 1994," NASA RP-1404, December 1997.

12 Thanks to Harold Saive for retrieving the manual from government microfiche: chemtrailsplanet.files.wordpress.com/2013/03/chemtrails-chemistry-manual-usaf-academy-1999.pdf

13 Thomas was carefully watched as early as July 9, 1999, the day he was to present "Chemtrails Over America: What's Wrong With Our Skies?" at the James A. Little Theater in Santa Fe, New Mexico, and two parallel trails simultaneously appeared overhead from east to west. Three weeks before his presentation, the *New Mexican* had run its first attack on chemtrail "conspiracy theorists."

14 *Misinformation* is defined as unintentional wrong information and *disinformation* as the intentional "spin" of true information with wrong information.

> Then when Hatfield revealed the coke bust in his book *Fortunate Son*, Rove in turn revealed that Hatfield was an ex-con with a conspiracy-to-murder conviction. *60 Minutes* did a hatchet job on Hatfield, and all copies of the book were burned by the publisher. From then on, no respectable journalistic outlet would go near the Bush cocaine arrest story.[15]

Given all the deaths of scientists who have broken with their "blood oath" confidentiality agreements,[16] the anonymity that "insiders" often insist upon for their own safety is understandable but can also work against what they are revealing. Here are three positive examples and one less so.

Angels co-author Manning had a fortuitous encounter with a "Mr. X" whose revelations about Pentagon "maniacs" and Bernard J. Eastlund's HAARP patents led directly to Manning and Begich digging into the HAARP project unfolding nearby in Gakona, Alaska, and writing *Angels* to forewarn the public. Then in 2001 "Deep Sky" stepped forward, a senior air traffic control manager for the northeastern seaboard of the United States who in three taped interviews with Will Thomas and veteran radio journalist S.T. Brendt insisted that "he had been ordered to divert incoming commercial flights away from USAF tankers spraying a substance that showed up on ATC radars as a 'haze'."[17]

Both early whistleblowers seemed authentic, as did the anonymous "well-placed military source" that independent scientist and Morgellons researcher Clifford E. Carnicom encountered in 2003. His insights into the biological aspect of the chemtrails aerosol operation were for the most part credible, not so much for who he was as for confirming other accounts from random angles:

> 1. The [aerosol] operation is a joint project between the Pentagon and the pharmaceutical industry.
>
> 2. The Pentagon wishes to test biological diseases for war purposes on unsuspecting populations. It was stated that SARS [severe acute respiratory syndrome] is a failure as the expected rate of mortality was intended to be 80%.
>
> 3. The pharmaceutical industry is making trillions on medications designed to treat both fatal and non-fatal diseases.
>
> 4. The bacteria and viruses are freeze-dried and then placed on fine filaments for release.
>
> 5. The metals released along with the diseases heat up from the sun, creating a perfect environment for the bacteria and viruses to thrive in the air supply.

---

15 Dr. Kevin Barrett, "Now It Can Be Told: The REAL Reason Obama Was Nearly Devoured by Carnivorous Plesiosaurs on Mars." *Veterans Today*, October 1, 2012.

16 Many scientists have fallen prey to untimely deaths; see "Dead Scientists 2004–2013," www.stevequayle.com/index.php?s=146.

17 William Thomas, "Air Traffic Controllers Concerned Over Chemtrails." Rense.com, March 4, 2002.

6. Most countries being sprayed are unaware of the activities and they have not consented to the activities. He states that commercial aircraft flying is one of the delivery systems.

7. Most of the "players" are old friends and business partners of the senior Bush [George H.W. Bush].

8. The ultimate goal is the control of all populations through directed and accurate spraying of drugs, diseases, etc.

9. People who have tried to reveal the truth have been imprisoned and killed.

10. His final words: "This is the most dangerous and dark time that I have experienced in all of my years of serving this country."[18]

But then the following year, the anonymous insider "Deep Shield"[19] showed up on the Internet to stir the pot, claiming that he worked with "chiefs and levels above the worker." Borrowing heavily from the National Academy of Sciences' 1992 *Policy Implications of Greenhouse Warming: Mitigation, Adaptation, and the Science Base,*[20] Deep Shield's revelations read chapter and verse like much of the conspiracy-spun pablum that dominated the first few years of the global Internet. Deep Shield hung heavy on chemtrails as a *Star Wars*[21] "shield" straight from the Strategic Defense Initiative (SDI) introduced in 1983 by the Reagan administration. (Under President Clinton, SDI was renamed the Ballistic Missile Defense and faded from headlines.)

Deep Shield insisted that the chemtrails agenda was about creating ozone in the stratosphere, stemming infrared and ultraviolet radiation, and terraforming a dying planet, and that public secrecy was merely to keep people from panicking. He claimed that defending against UV poisoning had brought the U.S., Russia, and NATO together and that they were all involved in chemical production and distribution, with observers on hand when aerosol canisters were being loaded. Because the "shield" had to be maintained for the sake of the entire planet, "nonparticipant nations" were being sprayed, too, and that *not* spraying was considered a military offense. He agreed that the tons of heavy metals and polymers being sprayed were making people sick, but insisted that "without spraying, we have a 90%+ chance of becoming extinct as a species within the next 20 years."[22] *Not once* did Deep Shield mention HAARP. Then, having spun his Internet spin, he dropped from sight with a suicide story trailing behind him.

---

18 Clifford E. Carnicom, "A Meeting," July 26, 2003, www.carnicominstitute.com

19 Chris Haderer and Peter Hiess quote "Deep Shield" in their 2005 book *Geoengineering, Chemtrails, and Climate as Weapon.* (See Chapter 3.)

20 National Academy of Sciences, Committee on Science, Engineering, and Public Policy, National Academy Press, Washington, D.C., 1992.

21 1977 George Lucas film. Note "Disney buying Lucasfilm, will release new 'Star Wars' film in 2015," *NBCNews,* December 2012.

22 www.holmestead.ca/chemtrails/shieldproject.html

Mainstream media play an essential role when it comes to silencing voices attempting to expose secret programs that constitute a danger to the public (*and* republic). Whether those dangers are nonthermal radiation, GMO foods, DU weapons, HAARP, or chemtrails loaded with ionized metallic particulates and pathogens, going public in the land of the National Security Act is risky, indeed.

In 2000, New Zealand environmental scientist Neil Cherry, Ph.D., gave damning testimony regarding electromagnetic radiation before the New Zealand Parliament and the European Parliament in Brussels; the next year, he was diagnosed with a fast-acting motor neuron disease and became increasingly immobile, dying two years later at 56.[23] New York attorney Andrew Campanelli[24] advised city councils across the United States about writing more protective telecom ordinances and suing to stop the onslaught of cell phone towers. One summer day, his new office landlords turned off the air conditioning and stairwell lighting, the elevator cable snapped and left him suspended between floors, and when he tried to use the elevator phone, he was cut off even as he smelled smoke. By the time the firefighters arrived, he'd crawled out onto a chest-high floor.[25]

The prominent German newspaper *Suddeutsche Zeitung*[26] reports that individuals and groups opposed to genetically modified organisms (GMOs) and the glyphosate in Monsanto's Roundup herbicide have been threatened, hacked, slandered, and terrorized. As Chapter 6 will reveal, when it comes to biotechnology and control over food production, geoengineering and megacorporations like Monsanto work in tandem. Monsanto has ties to U.S. intelligence, the military, and private security companies, and former Monsanto executives hold scores of key government positions.[27] While anti-chemtrail activists around the world (like Label GMO) attempt to awaken public awareness through protests like the Global March Against Chemtrails and Geoengineering (August 25, 2013; January 25, 2014; etc.), Monsanto consolidates its geoengineering clout by purchasing Climate Corporation, a farmer insurance underwriter, for $1 billion.[28]

According to a report prepared for Prime Minister Vladimir Putin by the Foreign Military Intelligence Directorate (GRU),[29] John P. Wheeler III, 66—

---

23 Dr. Neil Cherry, "Evidence that Electromagnetic Radiation is Genotoxic: The implications for the epidemiology of cancer and cardiac, neurological and reproductive effects," June 2000.

24 www.campanellipc.com/attorneys.htm

25 See anticelltowerlawyers.com, if it hasn't been hacked.

26 sustainablepulse.com/2013/07/13/the-sinister-monsanto-group-agent-orange-to-genetically-modified-corn/#.Uj92AeB9mrI

27 Jonathan Benson, "Groundbreaking investigation reveals Monsanto teaming up with US military to target GMO activists." *Natural News*, July 29, 2013.

28 "Monsanto Buys Weather Climate Corporation For 1 Billion," www.minds.com

29 "Top US Official Murdered After Arkansas Weapons Test Causes Mass Death." *European Union Times*, January 4, 2011. As usual, readers will have to make their own decisions regarding the veracity of this foreign source. Also, see Sarah Hoy, "Central Arkansas growing weary of relentless tremors." *CNN*, December 28, 2010: 500 measurable earthquakes in central Arkansas between September and December 2010. What all was going on in Arkansas?

a West Point, Harvard, and Yale insider—was brutally murdered on New Year's Eve 2010 after threatening to expose the U.S. military's test of the poison gas Phosgene that killed hundreds of thousands of animals in Arkansas. From 2005 to 2008, Wheeler was Special Assistant to the Secretary of the Air Force, then Special Assistant to the Acting Assistant Secretary of the Air Force for Installations, Logistics and Environment. He then worked as a consultant for MITRE Corporation, producer of the first draft of HAARP's Environmental Impact Statement, which bypassed congressional scrutiny and was instead announced in the Federal Registry.

Strange deaths happen to NASA employees, as well. Paul Milford Muller, 76, co-navigator on the Apollo Navigation Team, was found in Thailand in May 2013 with a rope around his neck and genitals, plus drug paraphernalia nearby. He had worked at NASA's Jet Propulsion Laboratory at the California Institute of Technology. Seven years earlier, former NASA astronaut Charles E. Brady, Jr., MD, 54, was found dead by knife wounds on Orcas Island, Washington. When chosen as an astronaut, Dr. Brady had been serving in the U.S. Navy's Tactical Electronic Warfare Squadron 129. He spent 16 days, 21 hours, 48 minutes and 30 seconds 173 miles in space above the Earth on space shuttle *Columbia* (STS-78), saying of his journey:

> You almost feel guilty being up there. The oceans were so blue, they were almost phosphorescent. The landmasses were something like you'd dream of as a child. The earth is a magnificent place, and it does look perfect from space. At the same time, we saw evidence of rain forest destruction and damaged river systems. It makes you come back feeling very deeply that you want to protect the earth.[30]

He added that he was totally unprepared for the almost corporeal transformation that takes place in space: "It is as if God suddenly made you into a dolphin. It is a mystery that is billions and trillions of times deeper than I had ever imagined."

How different from the military mindset that has loosed HAARP and chemtrails technology on Planet Earth.

3

Theoretical physicist Albert Einstein wisely said, "We cannot solve our problems with the same thinking we used when we created them." Learning new ways of thinking requires time, attention, practice, and a diligent, organized equanimity of mind—certainly not how prestigious Ph.D.s are trained to work on need-to-know research projects minus how their work will contribute to the whole picture. Thus is it any wonder that the scientific peer review system operates more like a good-old-boy club than an honest tool of oversight?

---

30 *The Pilot.*

As early as 1999, approximately 160 websites were already devoted to unlocking the chemtrail phenomenon. Today, the electronic buzz of thousands is furiously pursuing what is going on over our heads and how it connects with what is showing up in our soil and bodies. Researchers are slowly piecing together bits and pieces of evidence scattered throughout the media like Osiris' body—individual observations, analyses of Doppler maps, patents, scientific papers, military studies, etc.—all of which must be retrofitted into a picture of the military operations we have been denied access to. It is while thinking our way through a hunt-and-peck comparison of sources that we learn to think in ways that differ from the thinking that created the problem.

Meanwhile, mainstream media work hard to sustain military-industrial complex thinking. Take, for example, *Daily Comet* reporter Xerxes A. Wilson's 2012 article about the tremors in Lafourche and Terrebonne parishes, Louisiana, land of Big Oil fracking,[31] salt domes, and sinkholes. (See Chapter 5.) Wilson chooses descriptors like "vibration," "rumble," "shake," "loud noise similar to thunder," "sounding like a garbage truck had just dropped a dumpster," "more tremors," but when it comes to what might have caused the tremors, he is stone silent. He quotes the Lafourche Parish Sheriff's Office and Office of Emergency Preparedness regarding weather and flyovers, and the Geological Survey's National Earthquake Center regarding seismic activity, but references only hearsay when it comes to connecting the tremors with the sinkhole, and then not a word about fracking:

> Residents of the Assumption Parish community of Bayou Corne felt light tremors in the months leading up to the emergence of the 400-foot-wide sinkhole in the swamp near the community this past summer. Scientists believe the sinkhole was caused when a subterranean brine cavern collapsed within the Napoleonville Salt Dome. The floor of the cavern is more than 1,000 feet underground. Though Lafourche and Terrebonne have similar brine caverns, [the director of Emergency Operations] said there is no reason to believe there are any similarities to the Bayou Corne situation.[32]

"Lay Scientist" Martin Robbins wrote a clever piece in *The Guardian* about "opaque references to 'experts' and 'scientists' who 'say' or 'claim' things," and how referring to individual experts and bits of research "imply consensus where none exists" and mean "approximately jack shit."

> . . . newspapers train their readers to view science as a homogeneous community of 'them' and wonder why 'they' make apparently contradictory

31 Slang for hydraulic fracturing of rock to get more gas and oil to flow out.

32 Xerxes A. Wilson, *Daily Comet*, Lafourche Parish, Louisiana, October 4, 2012; also see Deborah Dupre, "Sinkhole state-ordered fracking-type process might be causing quakes." Examiner.com, October 27, 2012.

claims at different times . . . Stories based on single studies or experts are not unlikely to be wrong, and over-reliance on stock phrases like 'scientists say' suggests the writer hasn't grasped the wider picture. In the world of science, context is king — we need more of it.[33]

In 2010, "scientists" were "baffled" by the collapse of the thermosphere 56 miles above the Earth surface, then relieved when it "rebounded." NASA and the Naval Research Lab called it a "Space Age record."[34] Did these scientists lack HAARP need-to-know clearance, or were they just providing cover?

Will Thomas recounted a possible need-to-know HAARP anecdote in *Chemtrails Confirmed* (2004). In 1989 or 1990, Air Force pilots were invited to attend a NASA-sponsored meeting at NOAA in Washington, D.C. to toss around ideas about how large air masses might be moved to enlarge the launch window at Cape Canaveral. When a team of pilots offered the feedback that it didn't seem feasible, they were dismissed and discussion continued without them.

Daily hearsay accounts of inexplicable wonders may be true, and then again may be planted to obfuscate, distract, or create anxiety. What exactly were those "sky sightings" from Raton, New Mexico to Colorado Springs, Colorado and Liberal, Kansas on summer solstice 2012? And six months later on the last day of 2012, what caused hundreds of in-flight starlings to simultaneously drop dead southeast of Knoxville, Tennessee?[35] Aircraft battling blazes were grounded after a heavy tanker making a slurry run on the blaze spotted a "meteor." Was it a meteor, or was it a UFO or plasma orb? But no one's talking.[36] In Scotland, Alastair MacMillan saw a "ball of fire with a long green tail" 150–200 feet off the ground. "No whoosh, no squeak, just this bright, bright light," MacMillan said. Brian Guthrie added that he saw something large breaking up in the atmosphere. "I've seen shooting stars and meteor showers before, but this was much larger and much more colourful," said Guthrie.[37] Space junk? Satellite debris? Plasma orb?

Then there are the news sources themselves. Who funds them? Consider the *Veterans Today* story about a "highly unfriendly extraterrestrial threat."[38] Rumors of underwater bases in the Pacific Basin and advanced suborbital energy weapons being deployed from Vandenberg Air Force Base are one thing, but UFO wars? *Veterans Today* is a "military and foreign affairs journal," so what was senior editor Gordon Duff getting at? He ends the article with, "I either have to give this a 70% or reclassify a reliable official source as purposefully leaking

33 Martin Robbins, "Scientists say. . ." *The Guardian*, 6 March 2012.

34 John Emmert, Naval Research Lab, "Collapse in Earth's Upper Atmosphere Stumps Researchers." Tech Talk, *CBS News*, July 16, 2010.

35 "Birds fall from sky in Seymour." *The Mountain Press*, December 31, 2012.

36 "Meteor reports ground Colo. firefighting planes," *FOX News*, June 21, 2012.

37 "Meteor' sightings across Scotland prompt 999 calls." *BBC News*, 22 September 2012.

38 Gordon Duff, "UFO War: Chinese and US Navy off San Francisco: Joint Fleets Fend Off UFO Threat." *Veterans Today*, September 17, 2012.

something that makes no sense." Was it a cover story to obscure covert naval operations? Was Duff seeding cognitive dissonance[39]? Was the story true but its context false?

Electronic Internet articles and published papers are too easy to hack and "edit." Louis Slesin of *Microwave News* discovered that the White House Office of Policy Development had removed a key paragraph of a two-year Environmental Protection Agency (EPA) study, namely the recommendation that ELF fields be classified as "probable human carcinogens." Slesin's exposés are typically shunted to the fringe, as *Time* magazine finally noted in 1990:

> Like the one about the 23 workers at the Bath Iron Works in Bath, Me., who got "sunburns" one rainy day when someone on a Navy frigate flicked on the ship's radar. Or the trash fires that start spontaneously from time to time near the radio and TV broadcast antennas in downtown Honolulu. Or the pristine suburb of Vernon, N.J., that has both one of the world's highest concentrations of satellite transmitting stations and a persistent — and unexplained — cluster of Down's [sic] syndrome cases.[40]

Renaming clandestine projects and corporations that have received too much public attention is a widely employed shell game. For example, the High Accuracy Radial Velocity Planet Searcher, a radial velocity spectrograph installed in 2002 at ESO's 3.6-meter telescope at La Silla Observatory in Chile, uses the acronym HARPS from which the V for velocity has been deleted. Perhaps to properly confuse HARPS with HAARP, especially for a Google search?

If you have to be this cunning to ferret out a straight answer from news sources, imagine what is going on at the scientific research level where scientists "not formally tied to" defense patrons are rewarded with glorious careers, and scientists of conscience are either marginalized or silenced for good. For example, a decade ago, the international weekly *New Scientist* complained that the National Academy of Sciences (NAS) and the Pentagon's Joint Non-Lethal Weapons Program (JNLWP) refused to release dozens of reports on non-lethal weapons. Were chemtrails at issue?

> The Academy is justifying its unprecedented reticence by citing security concerns after 11 September. But campaigners think the real reason is that the research violates both U.S. law and international treaties on chemical and biological weapons . . . In 2000, *New Scientist* revealed that senior officials in the JNLWP want to rewrite the chemical and biological weapons treaties to give themselves more freedom to develop

39 Cognitive dissonance: the feeling of discomfort that follows from simultaneously holding two or more conflicting ideas.

40 Philip Elmer-Dewitt, "Technology: Hidden Hazards of the Airwaves." *Time*, July 30, 1990.

non-lethal weapons. The reports make it clear that research that violates the treaties has been underway since the 1990s.[41]

Finally, a word about the overused "C" slur. Robert Shetterly's clever response in "The Necessary Embrace of Conspiracy":

> *Con + spirare*, from the Latin. *To breathe together.* Those are the roots of conspiracy. Breathing together doesn't sound like an activity of the ideologically deracinated whispering seditiously in a dank cellar or a board room, foul breaths denting a weak flame flickering over a candle nub, gunpowder or greed blackened fingers setting a timer, the whites of creased eyes glinting like knives with treason, murder, power, and deceit. *Con + spirare* sounds like healthy men and women standing in the sun figuring out how in the hell they are going to take care of each other and their aging mother Earth and love life while doing it. Breathing together, sharing the same air, plotting to make sure that what's mine is yours, conspiring to save their self-respect, their ideals, the future for their children.
>
> I want to be part of a conspiracy. Pervasive, populist, revolutionary, and totally transparent. Grassroots. Idealistic. Simplistic. Life-affirming. Community-building.
>
> A conspiracy to make the common good and the love of nature the common denominator of every economic transaction.
>
> And the simple truth is either we start breathing together, conspiring big time, right out in the open, nakedly, unashamedly, or we will have conspired in secret, by default, in our own demise.
>
> We have let *them* breathe for us, and they have stolen our breath, our air, our spirit.
>
> Secret *con + spirare* is death. Open *con + spirare* is life.
>
> Conspiracy is dead. Long live conspiracy![42]

4

Thus far, we have covered the secrecy and silence that inquiries into truth must now labor under; the danger to insiders and anonymous sources whose conscience demands they speak out, as well as the danger of disinformation undermining researchers' credibility; and the nation's overarching burden of an embedded mainstream media that no longer serves the First Amendment.

---

41 Debora MacKenzie, "US non-lethal weapon reports suppressed." *New Scientist*, 09 May 2002. *New Scientist* seems to take chances: In January 2009, its cover article "Darwin was wrong" detailed how Darwin's evolution theory was skewed, one example being the shape of phylogenetic trees of interrelated species. Some evolutionary biologists boycotted the magazine.

42 Robert Shetterly, "The Necessary Embrace of Conspiracy." *Common Dreams*, August 31, 2007.

A good introduction should also offer readers a map by which to navigate the chapters ahead, especially when those chapters seek to throw light on a shadowy technology whose glaring absence from public debate threatens public safety. *Chemtrails, HAARP, and the Full Spectrum Dominance of Planet Earth* exposes how chemtrails (persistent contrails / aerosols) and ionospheric heaters like HAARP work together in a pump-and-dump action seeking to "own the weather,"[43] maintain a global wireless Smart Grid, and experiment with chemicals and biologicals on unaware populations.

The acronym "HAARP" will not only refer to the Ionospheric Research Instrument (IRI) in Alaska, nor to its several component installations at the High-Power Auroral Stimulation Observatory (HIPAS), the Chatanika Incoherent Scatter Radar Facility, and the Poker Flat Research Range. "HAARP" will also reference the global network of phased array ionospheric heaters and radar installations that together constitute a world *interferometric grid*. While HAARP's transmitter power may be only 3.6 MW (millions of watts),[44] its effective radiated power (ERP) can actually be multiplied to *billions* of watts (gigawatts):

> [The Gakona facility] has 180 antenna units, organized in 15 columns by 12 rows, yielding a theoretical maximum gain of 31 dB. *A total of 3.6 MW of transmitter power will feed it, but the power is focused in the upward direction by the geometry of the large phased array of antennas which allow the antennas to work together in controlling the direction* . . . The facility officially began full operations in its final 3.6 MW transmitter power completed status in the summer of 2007, *yielding an effective radiated power (ERP) of 5.1 Gigawatts* or 97.1 dBW [decibel watts] at maximum output.[45] [Emphasis added.]

In other words, by locking crossbeams with other ionospheric heaters, HAARP reveals itself to be a weapon of global scale.

The first section of *Chemtrails, HAARP, and the Full Spectrum Dominance of Planet Earth* is entitled "Ionizing the Atmosphere."

Chapter 1, "A Thumbnail History," offers an historical overview of the Tesla weapon now known as HAARP as well as a chronology of chemical and biological warfare experimentation that nonconsensual civilians and enlisted military personnel have been subjected to since the Cold War.

Chapter 2, "Deconstructing Eastlund's 1987 HAARP Patent," examines the HAARP patent and how the ionosphere is utilized as a "mirror" so beam weapons can, with enough charged ionospheric energy, employ phase interference that

---

43 See Appendix D, "Weather as a Force Multiplier: Owning the Weather in 2025," *Air Force 2025*, August 1996.

44 www.darpa.mil/ucar/programs/haarp.htm, March 9, 2006.

45 Wikipedia, "High Frequency Active Auroral Research Program."

can focus, localize, and pulse vast quantities of extra low frequency (ELF) energy on regions far from their sources.

Chapter 3, "Not Your Average Contrails," studies the aerosol delivery system known as chemtrails (as distinguished from contrails[46]) essential to the HAARP network. The nanoparticulates of conductive metals and polymers, sensors, microprocessors, and biological pathogens loaded into the chemtrails serve multiple agendas, the most crucial being to perpetuate a continuously ionized atmosphere.

Chapter 4, "The Poisons Raining Down," examines what jet fuel is dropping on us, along with a varying cocktail of particulate matter ("chaff"), polymers, mold and fungi, heavy metals, and sulfur, all acting synergistically, all challenging our immune systems.

The next section, "Profit and Force Multipliers," looks at everything from what Big Oil and HAARP tomography are doing to the Earth, to the biologicals and "cross-domain bacteria" taking up residence in our bodies.

Chapter 5, "Exploiting Earth Changes," comes to terms with "global warming" and looks closely at what really lies behind Earth events unfolding, from the BP oil spill in the Gulf of Mexico, subsidence, salt domes, and sink holes, to the earthquakes occurring along the New Madrid Seismic Zone.

Chapter 6, "Climate Engineering, Food, and Weather Derivatives," pinpoints the profit-and-plunder opportunity that HAARP-chemtrails "geoengineering," "Earth changes," and "extreme weather" present for disaster capitalists and corporate control over global food.[47]

Chapter 7, "Geoengineering and Environmental Warfare," examines weather engineering as a military "force multiplier" and treats the HAARP-chemtrails pump-and-dump as the Tesla weapon it is. Readers are offered a general idea of how this weapon works, followed by accounts of several major earthquakes, hurricanes, and tornadoes that appear to have been engineered as "force multipliers."

Chapter 8, "Morgellons: The Fibers We Breathe and Eat," concentrates on the hard research achieved by independent scientist Clifford E. Carnicom that connects the bioengineered fibers falling from chemtrails with the "nanowires" poking out of Morgellons sufferers' sores. I concentrate on Carnicom's work over that of other inquiring scientists for two reasons. The first is the sheer scope of his research on a painstaking budget that readers can easily follow at www.carnicominstitute.org. He has been studying the skies over his New Mexico home since the 1990s. The fact that his experiments have not been independently replicated by agencies like the Environmental Protection Agency

---

46 According to U.S. Patent #8,402,736 B2, "Method and apparatus for suppressing aeroengine contrails" (2010), chemtrails *and* contrails may soon be invisible.

47 See G. Edward Griffin, Michael J. Murphy, Peter Wittenberger, *What in the World Are They Spraying?* Truth Media Productions, 2010; and Michael J. Murphy and Barry Kolsky, *Why in the World Are They Spraying?* Truth Media Productions, 2012. All appendices can be found at feralhouse.com/chemtrails-appendices/

(EPA) and Centers for Disease Control and Prevention (CDC) is not his fault as these agencies, to which he has appealed again and again, have refused his requests or ignored them. Take the time to examine his papers and you will see why the National Security State "visitors" on the list at the end of this book continue to check on his work while marginalizing it. The second and lesser reason I concentrate on Carnicom is that he has been readily accessible for discussions and questions, given that we were friends long before I was asked to write this book. Destiny (or whatever you care to call it) has its own methods of getting things done.

By and large, I have pieced together data from open Internet sources and hearsay of acute observers. I hope to stimulate dialogue while expecting more qualified researchers to correct my errors. The few layman books on HAARP and chemtrails (see Bibliography and Footnotes) provided me with an excellent overview but are now dated, given more recent revelations regarding the cutting-edge disciplines of plasma physics, scalar interferometry,[48] nanotechnology, genomics, etc. Until then, the Internet remains the last vestige of a free press, despite its flaws and ongoing surveillance.

Chemtrails, HAARP, and the Full Spectrum Dominance of Planet Earth should be viewed as a learning manual, a work in progress for the sake of maintaining a biological (not to mention *human*) future, given that chemtrails-HAARP technology—global Tesla weaponry—is not just a scientific issue but a *moral* issue superseding "national security." It is with this in mind that I close with a fervent appeal written by Clifford Carnicom in 2004. May his words begin to dispel once and for all the belief in the cover story of greenhouse gases being the sole cause of "global warming":

> It can be demonstrated that the introduction of essentially any metallic or metallic salt aerosol into the lower atmosphere will have the effect of heating up that lower atmosphere. The impact is both significant and measurable. Those that seek and express concern on the so-called global warming problem might wish to begin their search with an inquiry into the thermodynamics of artificially introduced metallic aerosols into the lower atmosphere . . . An examination of the specific heat characteristics of an altered atmosphere will provide the path for the realistic conclusions that can be made.
>
> Any claim that the aerosol operations represent a mitigating influence on the global warming problem appears to be a complete façade that is in direct contradiction to the fundamental principles of physics and thermodynamics. The lack of candor and honesty by government, media and environmental protection agencies in response to public inquiry is further evidence of the

---

48 See Glossary for "Interferometry" and "Scalar"; also, see Appendix K "Scalar Wars: The Brave New World of Scalar Electromagnetics" by Bill Morgan (2001).

> fictitious fronts that have been proposed. It is past time to recognize that one of the primary effects of the dense aerosols that now permanently mar the lifeblood of this planet is the heating up of the very atmosphere we breathe.[49]

With this clarion call resonating throughout our electrical bodies and brains, we begin our study of chemtrails and HAARP and their role in what the military hopes will constitute a full spectrum dominance of Planet Earth and its living beings.

49 Clifford E. Carnicom, "Global Warming & Aerosols," January 23, 2004.

IONIZING THE ATMOSPHERE

This is America's gift to warfare.

— Admiral William Owens, vice chair of the Joint Chiefs of Staff (1994–1996)

It is not rational to assume that wars are always fought only the way we know them . . . We must permit ourselves to think of war as a silent, unnoticed, slow working but deliberate destruction of life on a planet, a satellite, or even a star. We must be ready to change every single view we have held sacred if it contradicts new facts and new experiences.

War may be going on right now with no one being aware of it, with men dying, with trees bending like rubber hoses, green pastures turning into dust bowls, and with academic and civil institutions explaining it all away as with 'just this' or 'just that.' In short, it may turn out correct what one would otherwise feel inclined to ascribe to a schizophrenic mind — namely, that instead of shooting at the victims of war with bullets, one could very well sap life energy out of the war victims with machines which operate according to the orgonomic potential of the Cosmic Energy.

— Wilhelm Reich

CHAPTER ONE

A Thumbnail History

Should a democratic people cede to its government the full responsibility of determining when secret tests on unwitting subjects are necessary to protect the nation's security?

— Senator Edward Kennedy, "Biological Testing Involving Human Subjects By the Department of Defense, 1977." Hearings before the Committee on Human Resources, 95th U.S. Senate

It will likely be years before Americans are told what is being tested upon them during our present chemtrail/space wars era. The Hanford downwinders did not learn until 1986 what had been unleashed upon them some 30 years earlier; SHAD [Operation SHAD, Shipboard Hazard and Defense] victims filed suit in 2003 to learn the extent to which they were intentionally exposed to dangerous substances in the '60s.

— Amy Worthington, "Chemtrails: Aerosol and Electromagnetic Weapons in the Age of Nuclear War," June 1, 2004

The weather balloons and sounding rockets of the Fifties gave way to the geosynchronous satellites and ionospheric heaters of the Seventies.

— Jerry E. Smith, *HAARP: The Ultimate Weapon of the Conspiracy*, 1999

At the outset of this study of chemtrails and their relationship to ionospheric technologies like the High-frequency Active Auroral Research Project (HAARP), it is crucial that we understand Section 1520a Chapter 32 of U.S. Code Title 50, "Restrictions on use of human subjects for testing of chemical or biological agents." Title 50 defines the role of war and national defense, and Chapter 32 sets limits on chemical and biological warfare programs. (See Appendix A.[1]) While the Secretary of Defense may not conduct any chemical or biological experiments on civilian populations, the loophole lies in allowing for medical, therapeutic, pharmaceutical, agricultural, and industrial research and tests, including research for protection against weapons and for law enforcement purposes like riot control.

1 All appendices can be found at feralhouse.com/chemtrails-appendices/

HAARP is classified as a research project, which means "informed consent" is suspended.

The history of HAARP's conception has been a work in progress over the past two decades in books, essays, and Internet research. (See Bibliography and Resources.) The at-a-glance timeline[2] from the 1995 *Angels Don't Play This HAARP: Advances in Tesla Technology* by Jeane Manning and Nick Begich will help to develop a frame of reference for just how long the Strategic Defense Initiative (SDI)—of which chemtrails and HAARP are a crucial, culminating part—has been under development. A narrative of key points from Dr. Rosalie Bertell's excellent "Background of the HAARP Project"[3] follows the Manning-Begich timeline. Dr. Bertell, a Grey Nun of the Sacred Heart for over half a century, was an epidemiologist best known for her ionizing radiation expertise. She died on June 14, 2012 at the age of 83. (See Appendix I.)

Angels Don't Play This HAARP Timeline

1886–8: Nikola Tesla invents system of Alternating Current power source and transmission system. As 60-pulse-per-second (Hertz) AC power grids spread over the land, Mom Earth will eventually dance to a different beat than her usual 7–8 Hertz frequency.

1900: Tesla applies for patent on a device to transmit "Electrical Energy Through the Natural Mediums."

1905: U.S. Patent #787,412 issued for above.

1924: Confirmation that radio waves bounce off ionosphere (electrically charged layer starting at altitude of 50 kilometers).

1938: Scientist proposes to light up night sky by electron gyrotron heating from a powerful transmitter.

1940: Tesla announces "death ray" invention.

1945: Atomic bomb tests begin 40,000 electromagnetic pulses to follow.

1952: W.O. Schumann identifies 7.83 Hertz resonant frequency of the earth.

1958: Van Allen radiation belts discovered (zones of charged particles trapped in Earth's magnetic field) 2,000+ miles up. Violently disrupted in the same year.

1958: Project Argus, U.S. Navy explodes three nuclear bombs in Van Allen belt.

1958: White House advisor on weather modification says Defense Dept. studying ways to manipulate charges "of earth and sky, and so affect the weather."

2 Yet another timeline and excellent discussion can be found at "Geoengineering Exposed: Global Warming Linked to Advanced Climate Modification Technology," chemtrailsplanet.net/tag/haarp

3 Rosalie Bertell, "Background of the HAARP Project." Earthpulse.com. See www.airforcebase.net/usaf/GWEN_list.html for a list of the 58 relay-node sites.

1960: Series of weather disasters begins.

1961 Project Westford: Copper needles dumped into ionosphere as "telecommunications shield."

1961: Scientists propose artificial ion cloud experiments. Beginning of dumping chemicals (barium powder, etc.) from satellites/rockets.

1961–62: Soviets and USA blast many EMPs in atmosphere, 300 megatons of nuclear devices deplete ozone layer by about 4%.

1962: Launch of Canadian satellites and start of stimulating plasma[4] resonances by antennae within the space plasma.

1966: Gordon J.F. MacDonald publishes military ideas on environmental engineering.

1960s: In Wisconsin, U.S. Navy Project Sanguine lays ELF antennae.

1968: Moscow scientists tell the West that Soviets pinpointed which pulsed magnetic field frequencies help mental and physiological functions and which do harm.

1972: First reports on "ionospheric heater" experiments with high-frequency radio waves at Arecibo, Puerto Rico. 100-megawatt heater in Tromsø, Norway built later in decade; can change conductivity of auroral ionosphere.

1973: Documentation that launch of Skylab "halved the total electron content of the ionosphere for three hours" (by rocket exhaust gases).

1973: Recommendations for study of Project Sanguine's biological effects denied by Navy.

1974: United Nations General Assembly bans environmental warfare.

1974: High-frequency experiments at Plattesville, Colorado, Arecibo, and Armidale, New South Wales heat "bottom side of ionosphere."

1974: Experiments' airglow brightened by hitting oxygen atoms in ionosphere with accelerated electrons.

1975: Stanford professor Robert Helliwell reports that VLF from power lines is altering the ionosphere.

1975: U.S. Senator Gaylord Nelson forces Navy to release research showing that ELF transmissions can alter human blood chemistry.

1975: Pell Senate Subcommittee urges that weather and climate modification work be overseen by civilian agency answerable to U.S. Congress. Didn't happen.

1975: Soviets begin pulsing "Woodpecker" ELF waves at key brainwave rhythms. Eugene, Oregon, one of locations where people were particularly affected.

---

4 *Plasma*: an ionized gas consisting of ions and free electrons; an electrically conductive gas.

1976: Drs. Susan Bawin and W. Ross Adey show nerve cells affected by ELF fields.

1979: Launch of NASA's [National Aeronautics and Space Administration[5]] third High-Energy Astrophysical Observatory causes large-scale, artificially induced depletion in the ionosphere. Plasma hole caused by "rapid chemical processes" between rocket exhaust and ozone layer . . . ionosphere was significantly depleted over a horizontal distance of 300 km for some hours."

1985: Bernard J. Eastlund applies for patent "Method and Apparatus for Altering a Region in the Earth's Atmosphere, Ionosphere and/or Magnetosphere." (First of three Eastlund patents assigned to ARCO Power Technologies Inc.)

1986: U.S. Navy Project Henhouse duplicates Dr. Jose Delgado (Madrid) experiment: very low-level, very low-frequency pulsed magnetic fields harm chick embryos.

1980s: In the later part of the decade, the U.S. begins network of Ground Wave Emergency Network (GWEN) towers, each to generate Very Low Frequency (VLF) waves for defense purposes.

1987–92: Other APTI scientists build on Eastlund patents for development of new weapon capabilities.

1994: Military contractor E-Systems buys APTI, holder of Eastlund patents, and contracts to build biggest ionospheric heater in world, euphemistically named *High-frequency Active Auroral Research Program* (HAARP).

1994: Congress freezes funding on HAARP until planners increase emphasis on earth-penetrating tomography uses for nuclear counter-proliferation efforts.

1995: Raytheon buys E-Systems and old APTI patents. The technology is now hidden among thousands of patents within one of the largest defense contractor portfolios.

1995: Congress budgets $10 million for 1996 under "nuclear counter-proliferation" efforts for HAARP project.

1995: HAARP planners to test Patent #5,041,834 in September.

1994–6: Testing of first-stage HAARP equipment continues, despite frozen funding.

1996: HAARP planners to test earth-penetrating tomography applications by modulating the electrojet at Extremely Low Frequencies.

From Dr. Bertell's "Background of the HAARP Project," we learn that in 1924, it was confirmed that radio waves bounce off of the ionosphere 48,000 to 50,000

---

5 NASA was co-opted by the U.S. military in April 1982, when Dr. Christopher Columbus Kraft, Jr. was fired as director of Johnson Space Center.

kilometers above the surface of the Earth. Thanks to Operation Paperclip[6], rockets and nuclear weapons developed in tandem between 1945 and 1963, both entailing intensive atmospheric testing (stratosphere and ionosphere) as well as thousands of detonations above and below the surface of the Earth. Of the following upper and lower atmospheric "geography" terms, the most essential to this book will be the troposphere, ionosphere, and magnetosphere:

Troposphere: sea level — 6–20 km (3–12 mi.)
Stratosphere [contains ozone layer]: 20–50 km (12–31 mi.)
Mesosphere: 50–80 km (31–49 mi.)
Thermosphere: 80–690 km (49–428 mi.)
Exosphere: 690–10,000 km (428–6,213 mi.)
Ionosphere: top of the mesosphere to the top of the thermosphere
Magnetosphere: outer layer of the ionosphere — 60,000 km (37,282 mi.)

Protective Van Allen Belts, discovered in 1958 by Explorer I, begin at about 7,700 kilometers and extend to 51,500 kilometers. Like a net, these Belts capture charged particles from solar and galactic winds and spiral them gently down along the magnetic lines of force that comprise the magnetosphere surrounding the Earth, until they converge at the poles to spin Arctic and Antarctic auroras. The Sun's radiation creates and maintains both the ionosphere and plasma in the magnetosphere.

In her book *Planet Earth: The Latest Weapon of War*, Dr. Bertell defines and discusses plasma in the context of weapons, communication systems, and "space shields":

> . . . plasmas are superheated gas and they occur naturally in the ionosphere. We can see the effect of plasma on an object when we consider how the ionized layer of the Earth's atmosphere burns up the space debris and meteorites that enter it. It is not really friction with the more dense atmosphere which causes the extreme temperature, but the impact of the spacecraft itself, which compresses the highly active plasma, causing its temperature to increase dramatically—it can even temporarily reach the temperature of the surface of the sun. The Space Shuttle has insulation tiles on its surface to protect it from this heat.[7]

Throughout the 1950s and 1960s, more than three hundred megatons of nuclear bombs were exploded in the atmosphere—charged and radioactive particles, 40,000 electromagnetic pulses. From 1949 to 1957, Strontium-90

6 Among the first books out about the half-century secret Nazi importation was Linda Hunt's *Secret Agenda: The United States Government, Nazi Scientists, and Project Paperclip, 1945 to 1990* (New York: St. Martin's Press, 1991).

7 Rosalie Bertell, *Planet Earth: The Latest Weapon of War*. The Women's Press Ltd., 2000.

increased in baby teeth 14-fold. During the Leonids of 1958, Project Argus exploded three nuclear fission bombs 480 km above the South Atlantic Ocean and created new magnetic radiation belts four to eight thousand miles above the Earth. Radioactive I-131 particles with a half-life of eight days settled in thyroid glands. Cows ate fallout grass, Americans drank fallout milk. Strontium-90, cesium-137, zirconium, and other radioactive isotopes seeped into living tissues.

In 1961 under Project Westford, 350,000 million copper needles 2–4 cm long were dumped 3,000 kilometers above the Earth to create an ionospheric telecommunications shield against magnetic storm and solar flare disruptions. Three satellites were destroyed, Hawaii blacked out and the sky over the Arctic Pole was ignited. As another radiation belt was formed at 400–1,600 km above, the inner Van Allen Belt was destroyed and billions of free electrons plummeted into our atmosphere ("electron shower"). In a similar manner, the Soviet Union created three more belts at 7,000–13,000 km.

On July 9, 1962, Project Starfish detonated three nuclear blasts in the upper atmosphere. As for what the one-kiloton, one-megaton, and one-multi-megaton devices wreaked high above the Earth, here is how the Keesings Historisch Archief (KHA)—an archive of factual and objective information about all major developments following World War II—reported it:

> . . . the inner Van Allen Belt will be practically destroyed for a period of time . . . [and] the ionosphere . . . will be disrupted by mechanical forces caused by the pressure wave following the explosion. At the same time, large quantities of ionizing radiation will be released . . . On 19 July . . . NASA announced that as a consequence of the high-altitude nuclear test of July 9, a new radiation belt had been formed, stretching from a height of about 400 km to 1,600 km [300–1,200 mi.] . . .[8]

Needless to say, the electron fluxes of the lower Van Allen Belt have never been the same. Too little too late, the 1963 international Limited Test Ban Treaty banned nuclear testing underwater, in the atmosphere and outer space—but not underground.

ANOTHER MANHATTAN PROJECT?

When the *Saturn V* rocket malfunctioned in 1975 and burned a large hole in the ionosphere that prevented all telecommunications over a large area of the Atlantic Ocean, interest in the ionosphere tripled. Ionospheric heating experiments were conducted in Plattesville, Colorado, Armidale, South Wales, Australia, and Arecibo, Puerto Rico. The Max Planck Institute built EISCAT, the European

8 *KHA*, 29 June 1962, 11 May 1962, 5 August 1962.

Incoherent Scatter Radar, a 48-megawatt heater in Tromsö, Norway now run by a five-country consortium said to be engaged in post-Wundtian experiments.[9]

The ionospheric heater lead-up to the Strategic Defense Initiative (SDI) "Star Wars" had arrived.

Initially, the 1968–1980 Solar Power Satellite Project (SPS) was sold to the taxpaying public as an energy program: solar-powered satellites in geostationary orbits 40,000 km above the Earth would collect solar radiation and transmit it via microwave beams back to Earth *rectennas* (receiving antennae) to provide for all U.S. energy needs by 2025 at $3,000/kW. Each satellite would be the size of Manhattan, and each rectenna site would take up 145 square kilometers of desolate, lifeless land here on the Earth's surface.

Unfortunately, this relay of "free energy" wouldn't be all that free. In fact, it was just another cover story for military objectives, rather like the explore-the-universe rockets to the moon. Satellites were about jamming or creating communications, surveillance, and beam weapons, and ionospheric heaters were about weaponizing the aurora electrojet, as the military likes to call it. Turn microwaves into giant virtual antennae, circuit boards, and low-frequency radio transmitters and they can be made to penetrate the ocean and Earth as well as impact satellite functions.

With ionospheric heaters and the SPS program up and running, experiments in fine-tuning jet stream diversions continued quietly. Eventually, ionospheric heaters could generate 3 million gigawatts (giga = billion) to heat up one square kilometer of the ionosphere to 28,000°C (50,432°F). Control the jet stream and you can reroute storms, hurricanes, and droughts and make a fortune from weather derivatives. Shoot microwaves into the Earth (tomography) and you can discover oil reserves, underground bunkers, and war booty. Decimate a developing country, then send in contractors, earth-moving machinery, "aid," CIA, military, etc. to boost the economy back home, and if a country does not comply with transnational and IMF "guidelines," a flood or drought might convince its leaders to reconsider.

In 1976, the National Weather Modification Policy Act ended up under the jurisdiction of the Secretary of Commerce. (See Chapter 5 for how lucrative weather derivatives are.) Then came the 1978 ENMOD Convention (Environmental Modification) prohibiting the deliberate manipulation of natural processes having severe effects on climate and weather patterns, such as earthquakes, cyclones, tidal waves, etc. Not only were both attempts to curtail the imminent weaponizing of weather toothless, but political leaders were given free rein to employ weather against their own people.

As a member of the SPS review panel under former President Carter (1977–1981), Dr. Bertell took her deep reservations to the U.N. Committee on Disarmament, but its helpless response was that as long as the project was

9 Wilhelm Wundt (1832–1920), creator of social or mass psychology.

classified as solar energy *research* and not as a weapons project, they could do nothing. HAARP too has been classified as a research project and "persistent contrails" as Solar Radiation Management (SRM)—not weapons.

In 1985, President Ronald Reagan, former CIA director and then Vice President George H.W. Bush, and Secretary of Defense Dick Cheney quietly moved the SPS Project from the Department of Energy (DOE) to the much larger budget of the Department of Defense (DoD). Now, the lead-up to HAARP began in earnest. By 1981, Orbit Maneuvering System space shuttles were injecting chemical gases into the ionosphere to induce ionospheric holes (deplete local plasma) by means of "very low frequency [VLF, ELF] wavelengths, on equatorial plasma instabilities, and on low frequency radio astronomical observations" (*Advanced Space Research*, Vol. 8, No. 1, 1988).

Throughout the 1980s, rocket and space shuttle launches averaged 600 per year, peaking in 1989[10] with 1,500 and many more during Gulf War I (2 August 1990–28 February 1991), all releasing hydrochloric acid and ozone-destroying chlorine into the stratosphere, with those launched since 1992 injecting even more. [Up near the Alaskan HAARP site, hundreds of rockets have been launched from Poker Flat Research Range alone; other fixed launch sites include Wallops Test Range (Virginia), White Sands Missile Range (New Mexico), Andoya Rocket Range (Norway), and Esrange (Sweden).]

Bertell closes her survey of 50 years of destructive programs targeting control of the upper atmosphere with, "The ability of the HAARP/Spacelab/rocket combination to deliver very large amounts of energy, comparable to a nuclear bomb, anywhere on earth via laser and particle beams, is frightening."[11]

Thus we see that geoengineering did not arrive out of the blue, so to speak. The crucial point, however, may be this: As the nuclear age was initiated under a cloak of stealth and deceit in the Manhattan Project, so the electromagnetic age is being initiated in the same way as "solar radiation management." Big science and big money (private and public[12]) have segued smoothly from nuclear to plasma physics. For one example among many, theoretical physicist Freeman Dyson, a key Manhattan Project player, has been a big Solar Radiation Management proponent. *Huffington* columnist Jay Michaelson even calls this ongoing covert operation a "non-regulatory 'Manhattan Project'":

> The projected insufficiency of Kyoto's emission reduction regime [sic], and the problems of absence, cost, and incentives discussed in part II, cry out for an

---

10 The same year that "two Black Brant X's and two Nike Orion rockets were launched over Canada, releasing barium at high altitudes and creating artificial clouds . . . observed from as far away as Los Alamos, New Mexico." Rosalie Bertell, "Background of the HAARP Project." Earthpulse.com. See www.airforcebase.net/usaf/GWEN_list.html for a list of the 58 relay-node sites.

11 Bertell, "Background of the HAARP Project." See Madison Ruppert, "German company demonstrates laser weapon capable of shooting down drones from over a mile away." *End the Lie*, no date.

12 HAARP decision-making pivots around Lawrence Livermore and Los Alamos National Labs, Kirtland and Wright-Patterson Air Force Bases.

> alternative to our present state of climate change policy myopia. Geoengineering—intentional, human-directed manipulation of the Earth's climatic systems—may be such an alternative. This part proposes that, unlike a regulatory "Marshall Plan" of costly emissions reductions, technology subsidies, and other mitigation measures, a non-regulatory "Manhattan Project" geared toward developing feasible geoengineering remedies for climate change can meaningfully close the gaps in global warming and avert many of its most dire consequences.[13]

HAARP and other ionospheric heaters around the globe listed later in this chapter do seem to represent another Manhattan Project.[14] Meanwhile, the production of "mini-nukes" and bunker busters gets a shot in the arm and the DOE resumes nuclear testing in Nevada even as it and the DoD are exempted from environmental laws.[15]

Of course, the electromagnetic age and all things HAARP must begin with Nikola Tesla, the naïve Serbian genius allowed to die in poverty and ignominy, after which the FBI "confiscated" his research. In 1940, Tesla announced his Death Ray, and in 1943 he was dead. As will be shown in Chapter 2, Eastlund's original 1987 HAARP patent #4,686,605 referenced two *New York Times* articles about Nikola Tesla (8 December 1915 and 22 September 1940). This should not be surprising, given that HAARP is a Tesla magnifying transmitter (TMT) whose secret, according to Eastlund, lies in its steering agility and pulsed transmissions. Lift part of the ionosphere and *you can make it move,* Eastlund said, *do things with it,* like generate one watt of heat per cubic centimeter (3.6 gigawatts), control weather by altering upper atmosphere wind patterns, change molecular compositions of specific regions of the Earth's atmosphere, conduct wireless power transmissions, knock out some radio communications while retaining others, create electromagnetic pulses, etc.

TESLA, THE ARCTIC, AND THE RUSSIAN WOODPECKER

Tesla understood the global electrical grid inscribed in the Earth and how the ground waves pulsed at an extremely low frequency (ELF) known as the Schumann resonance, which occurs "because the space between the surface of the Earth and the conductive ionosphere acts as a closed waveguide."[16]

Think of the global magnetic grid as the Earth's acupuncture meridians. As in acupuncture, what happens to the Earth's *qi* or *chi* affects all biological life

13 *Stanford Environmental Law Journal,* January 1998.

14 In the report on the 2009–2010 House of Representatives Committee on Science and Technology hearings "Geoengineering: Parts I, II, and III," the Manhattan Project was referenced three times.

15 [The Energy and Water Development Appropriations Act of 2004] Nick Anderson "House Approves Pentagon Wish List — Bill Includes Military Exemptions from Environmental Laws." *Los Angeles Times,* November 8, 2003.

16 "Flaring hot summer," www.sunlive.co.nz/news/37455-flaring-hot-summer.html

forms living within those meridians and the tectonic plates beneath them. Tesla knew that this grid was a repository of gravitational energy far more valuable than oil, and that the men bankrolling his experiments would want to use it not for free energy but for warfare. On April 21, 1908, Tesla wrote a letter to the editor of the *New York Times* about how his TMT technology could be made into a Death Ray:

> When I spoke of future warfare I meant that it should be conducted by a direct application of electrical waves without the use of aerial engines or other implements of destruction . . . This is not a dream. Even now wireless power plants could be constructed by which any region of the globe might be rendered uninhabitable without subjecting the population of other parts to serious danger or inconvenience.

Seventy days after his letter ran, an explosion 1,000 times more powerful than the Hiroshima atomic bomb took place near the Tunguska River in central Siberia (60 degrees 55'N 101 degrees 57'E). People and herds of Siberian reindeer were decimated, 500,000 acres of pine forest flattened. Just *before* the explosion, witnesses saw a glow in the sky for miles and yet no telescope detected incoming meteors or comet debris.

One week after the blast, Admiral Robert E. Peary, USN, set off from New York for the North Pole with 23 men, arriving in the spring of 1909 with only five of the original 23.

Historical accounts insist that no scientific investigation of Tunguska took place for 19 years, during which rumors of a meteorite circulated. When a scientific expedition finally arrived at the site in 1927, no impact crater was found, no nickel or iron to a depth of 118 feet. Dust and ice of Encke's comet were then trotted out, which is still what you will find if you go looking for what occurred at Tunguska.

Tesla's Wardenclyffe transmitter would have been capable of inducing such an explosion and more. In 1917, Tesla wrote, "When unavoidable, the [TMT] may be used to destroy property and life." Was he speaking from experience? For Jerry Smith, author of *HAARP: The Ultimate Weapon of the Conspiracy* (1998), this 1934 letter from Tesla to J.P. Morgan, Jr. alludes to an "experiment" like Tunguska:

> The flying machine has completely demoralized the world so much so that in some cities, as London and Paris, people are in mortal fear from aerial bombing. The new means I have perfected affords absolute protection against this and other forms of attack . . . These new discoveries I have carried out experimentally on a limited scale, created a profound impression.

Given the lingering controversy over whether or not Peary was ever at the Pole[17], was Peary's team always secretly heading for Tunguska? Did Tesla's magnifying transmitter overshoot his Pole coordinates? Could Tesla broadcast power explain what happened 80 years before HAARP?

Acutely aware of Tesla science by the late 1950s, the Soviets dreamed of giant geoengineering feats in the Arctic. One idea was to make the Arctic Ocean navigable and warm the far north of Siberia and expand agricultural *akp* (acreage) by damming the Bering Strait: pump water from the Arctic Ocean into the Pacific and draw warm water northward from the Atlantic to melt the polar ice pack. Another plan in the mid-1970s was to build a dam across the Bering Strait and *use atomic reactors to heat the oceans.* If things got *too* hot, they could always use meteorologist Henry Wexler's idea of launching a ring of dust particles into equatorial orbit—an idea remarkably similar to chemtrails loaded with particulates.

Warming the globe, not cooling it, was what the headlines were wringing their hands over in the early days of the Cold War between Russia and America. One Wexler/Russian idea to increase global temperature by 1.7°C (35.06°F) was to inject a cloud of ice crystals into the polar atmosphere by detonating ten H-bombs in the Arctic Ocean.[18] Or spray several hundred thousand tons of chlorine or bromine from a stratospheric airplane and destroy the ozone to abruptly increase the surface temperature of the Earth. Or dump metallic dust or CO_2 into the atmosphere to enhance the "greenhouse effect" and hence warm the planet. Some military planners even tossed around the idea of putting giant mirrors or unshielded nuclear reactors in orbit around the planet to act as "new suns"—all of which sounds uncomfortably familiar.

In 1957, scientists Roger Revell and Hans Suess co-authored a paper about the "greenhouse effect," suggesting that oceans would absorb excess carbon dioxide at a much slower rate than previously predicted.[19] The bonus was that carbon dioxide was warming the climate for free, so expensive and risky geoengineering projects wouldn't be needed. But the debate in the media would soon change from how to warm the Arctic to how to avoid warming the Arctic.

In 1962—the year Project Starfish blew the nuclear devices that created a band of trapped electrons in the force lines of the magnetosphere—Wexler was researching the link connecting chlorine and bromine compounds to the destruction of the stratospheric ozone layers. Just before giving a lecture entitled "The Climate of Earth and Its Modifications" at the University of Maryland Space Research and Technology Institute, 51-year-old Wexler died of a heart attack at Woods Hole, Massachusetts. One year after his untimely

---

17 See "The Cook-Peary North Pole Dispute," polarflight.tripod.com/controversy1.htm

18 Wexler, H. "Modifying Weather on a Large Scale," *Science*, Oct. 31, 1958.

19 Roger Revell and Hans Suess, "Carbon dioxide exchange between atmosphere and ocean and the question of an increase of atmospheric CO_2 during the past decades." *Tellus*, 1957.

death, the flip-flop from warming the Arctic to cooling the Arctic was made official. As meteorologist James Rodger Fleming put it, "Had Wexler lived to publish his ideas, they would certainly have been noticed and could have led to a different outcome and perhaps an earlier coordinated response to the issue of stratospheric ozone depletion."[20] Wexler's only epitaph is a crater on the Moon named after him.

History claims that Wexler's proposals were the last of the Cold War proposals to warm the Arctic, but they weren't. Philip L. Hoag, author of *No Such Thing As Doomsday: How to Prepare for Power Outages, Terrorism, War, and Other Threats* (1999), maintains that the same U.S.-Soviet cooperation that led to the Russian Woodpecker (see below) actually began in 1971 with POLEX (Polar Experiment of the Global Atmospheric Research Program) and continued in 1973 with AIDJEX (Arctic Ice Dynamics Joint Experiment). Both were exploring the possibility of melting the polar ice cap.[21]

In 1975, the Soviets created their own Tesla Magnifying Transmitter (TMT) in the ELF Woodpecker grid at Daga3. In 1982, Russian scientist Victor Sedletsky explained that the Woodpecker was an entirely new RADAR technology that would one day control any place on the globe. By "control," he meant *brain* control, given that the brain is a natural receiver. The Woodpecker network of transmitters—Angarsk and Khabarovsk in Siberia, Gomel, Sakhalin Island, Nikolayev in the Ukraine, Riga in Latvia, and a site 60 miles south of Havana, Cuba—is capable of creating a *psychotronic*[22] field or ELF scalar grid over the United States. According to nuclear engineer Thomas E. Bearden (Appendix J), the Woodpecker is based on Tesla gravitational waves of pure potential that Bearden calls *non-Hertzian scalar waves*. These ELF waves are transmitted in pairs *through the Earth* and converge at a predetermined latitude-longitude point on the Earth's surface.[23]

What is telling is that the U.S. government established Project Woodpecker[24] in 1977 and sold the Soviets the supermagnet[25] that put the TMT Woodpecker firmly in business:

> This magnet was a 40-ton monster capable of generating a magnetic field 250,000 times more powerful than that of the earth's magnetic field. Its purpose was to override, blank out, and/or interfere with the earth's natural

20 James Rodger Fleming, *Fixing the Sky: The Checkered History of Weather and Climate Control*. Columbia University Press, 2011.

21 Philip L. Hoag, "Weather Modification," 1996. *TWM Reference Index*, homepages.ihug.co.nz/~sai/ind3.html

22 The Russian term for parapsychology, brainwashing, and mind control.

23 Philip L. Hoag, "Weather Modification," 1996. homepages.ihug.co.nz/~sai/wxwar.html

24 www.sourcewatch.org/index.php?title=Project_Woodpecker. Lawrence Livermore National Laboratory was and is the main research center for U.S. development of the Project.

25 Perhaps an early particle accelerator or superconducting quantum interference device (SQUID).

> magnetic field to permit the Soviet Woodpecker to penetrate to the United States. The United States not only knew what it was for, they sent a team of scientists to help the Russians install it![26]

Given the pivotal U.S. role in fine-tuning the Woodpecker grid, I have to agree with Jerry Smith when he says, "HAARP seems to be proof that the United States joined the Soviets in attempting to duplicate Tesla's experiments. This could be the real explanation of why Tesla's name has disappeared from American history books . . ."[27]

As the Woodpecker's TMT signal was torqued up, standing waves pulsed through rock and earth, causing the molten core to resonate and earthquakes to increase. Insider scientist Andrija Puharich wrote in January 1978:

> Of the many great earthquakes of 1976, there is one that demands special attention—the July 28, 1976 Tangshan, China earthquake. The reason that this 1976 earthquake attracted my attention is that *it was preceded by a light flare-up of the entire sky over Tangshan.* Also, this earthquake occurred during the first month of Soviet Woodpecker radio emissions . . . The most prominent effect was that when the Soviet Woodpecker emission was on at full strength, the sky would light up like an ionized-gas lamp —*just as Tesla had predicted.*[28]

The telltale TMT sign indeed appeared at 3:42 A.M. over Tangshan, just as it had over Tunguska. The sky lit up like daylight; red and white lights were seen up to 200 miles away. Leaves were burned crispy, vegetables scorched on one side "as if by a fireball" (*New York Times,* June 5, 1977). Tangshan was destroyed, along with its 650,000 people.

Purportedly at NASA's behest, the National Academy of Sciences (NAS) created a weather modification program to cloak the classified military weather modification and warfare program[29], and it's been global warming ever since, despite professional corrections. On June 13, 2012, Fritz Vahrenholt, a German green energy investor and professor at the University of Hamburg, delivered the 3rd Global Warming Policy Foundation Annual Lecture at the Royal Society in London. Vahrenholt dismissed the UN's Intergovernmental Report on Climate

---

26 Jerry Smith, *HAARP: The Ultimate Weapon,* 1998.

27 Ibid.

28 Andrija Puharich, MD, LLD. "Global Magnetic Warfare: A Layman's View of Certain Artificially Induced Unusual Effects on the Planet Earth During 1976 and 1977," January 1978.

29 For insight into how *covertly* weather modification has been pursued, see two crucial reports: (1) Homer E. Newell, "A Recommended National Program in Weather Modification: A Report to the Interdepartmental Committee for Atmospheric Sciences," November 1966; and (2) "Weather Modification: The Evolution of An R&D Program into A Military Operation" (anonymous), a 1986 critique of the 1966 report. Both are on the Internet at this time.

Change (IRCC) edited by Greenpeace[30] as being littered with falsities and lacking in scientific data. Vahrenholt went on to add:

> Real, hard data from ice cores, dripstones, tree rings and ocean or lake sediment cores reveal significant temperature changes of more than 1°C, with warm and cold phases alternating in a 1,000-year cycle. These include the Minoan Warm Period 3,000 years ago and the Roman Warm Period 2,000 years ago. During the Medieval Warm Phase around 1,000 years ago, Greenland was colonized and grapes for wine grew in England. The Little Ice Age lasted from the 15th to the 19th century. All these fluctuations occurred before man-made CO_2 . . . The IPCC's current climate models cannot explain the climate history of the past 10,000 years. But if these models fail so dramatically in the past, how can they help to predict the future?[31]

Had alarmist assertions about temperatures rising to 4.5°C (40.1°F) by a magical doubling of CO_2 in the atmosphere been contrived? The argument that those who doubt the media hype about "global warming" and "climate change" are being manipulated by corporations that don't want to pay the carbon tax is thin at best, given that the same corporations are funding climate engineering, too. As Amy Worthington puts it, "The harebrained scheme of particle engineering was contrived to ensure that industry polluters will never be forced to decrease their greenhouse gas emissions . . . The warming mitigation program is a hoax . . ."[32]

Solar magnetic field patterns have lowered significantly in the past 150 years and will de-intensify in the decades ahead as global warming gradually reverses by 2030 to 2040. Susanne Posel closes her "Globalists Switching Gears" with, "Perhaps Vahrenholt's appearance at the globalist think-tank, the Royal Society, is some proof that [the global Elites] are beginning to alter their tactics."[33]

ENTER GEOENGINEERING AND BIG BUSINESS

Military projects have a tendency to eventually be "outsourced" or privatized to universities and defense contractors, a practice lending itself to cloaking and plausible deniability. Conveniently, space shuttles have been retired and "space travel" passed to for-profit corporations and wealthy men.

---

30 In "Globalists Switching Gears: Royal Society Lecturer Says CO_2 Not Affecting Earth's Temperature" (Activist Post, June 20, 2012), Susanne Posel calls Greenpeace "a UN propaganda arm disguised as a proponent of environmental concern." Posel is chief editor of *Occupy Corporatism.*

31 Fritz Vahrenholt, "Global warming: second thoughts of an environmentalist." *The Telegraph,* 18 June 2012.

32 Amy Worthington, "Chemtrails: Aerosol and Electromagnetic Weapons in the Age of Nuclear War." Globalresearch.ca, June 1, 2004.

33 Posel, June 20, 2012.

> . . . With the space shuttle's retirement Thursday, no longer will flying people and cargo up to the International Space Station be a government program where costs balloon. NASA is turning to private industry with fixed prices, contracts and profit margins. The space agency will be the customer, not the boss . . . NASA has hired two companies — Space Exploration Technologies Corp. of Hawthorne, Calif., and Orbital Sciences of Dulles, Va. — to deliver 40 tons of supplies to the space station in 20 flights. The cost is $3.5 billion, about the same price per pound as it was during the space shuttle's 30-year history.
>
> "It's time. Once NASA blazes the trail, creates the technology and it's available for private companies to take advantage of, this is the time" for the private firms to take over, said NASA commercial cargo chief Alan Lindenmoyer.[34]

With surveillance and laser satellites up and space militarized, it's time to turn it all into a business model and make some money—and bypass congressional oversight. This includes secret aerosol projects under the private sector canopy of geoengineering. In early 2012, *The Guardian* described it this way:

> A small group of leading climate scientists, financially supported by billionaires including Bill Gates, are lobbying governments and international bodies to back experiments into manipulating the climate on a global scale to avoid catastrophic climate change.[35]

What is not broadcast is the truth about what's causing "catastrophic climate change."

Other wealthy men besides Gates investing in geoengineering include founder and chair of the Virgin Group, Sir Richard Branson, tar sands magnate Murray Edwards, and co-founder of Skype Niklas Zennström—all backing "shield" geoengineers David Keith at Harvard University and Ken Caldeira at Stanford University,

> the world's two leading advocates of major research into geoengineering the upper atmosphere to provide the earth with a reflective shield. They have so far received over $4.6 million from Gates to run the Fund for Innovative Climate and Energy Research (Ficer) . . . According to statements of financial interests, Keith receives an undisclosed sum from Bill Gates each year, and is the president and majority owner of the geoengineering company Carbon Engineering, in which both Gates and Edwards have major stakes — believed to be together worth over $10 million . . .

---

34 Seth Borenstein, "NASA To Privatize Space Travel After Last Shuttle Lands." *Huffington Post*, July 20, 2011.

35 John Vidal, "Bill Gates backs climate scientists lobbying for large-scale geoengineering." *The Guardian*, 5 February 2012.

> "There are clear conflicts of interest between many of the people involved in the debate," said Diana Bronson, a researcher with Montreal-based geoengineering watchdog ETC. "What is really worrying is that the same small group working on high-risk technologies that will geoengineer the planet is also trying to engineer the discussion around international rules and regulations. We cannot put the fox in charge of the chicken coop."[36]

Still, putting foxes in charge of the chicken coop isn't new. In July 2012, Russ George, former CEO of Planktos, Inc., exercised his godlike powers to alter Nature by dumping one hundred tons of iron sulfate off the north coast of British Columbia. George claims he was attempting to stimulate phytoplankton bloom to create a carbon dioxide sink—a "sequestering" idea broached in 1988 by John H. Martin, an American oceanographer—but who knows what the wealthy foxes are really up to, given that Environment Canada, Canadian Space Agency, NASA, and NOAA unofficially "cost-shared" equipment and satellite resources with George.

> "It appears to be a blatant violation of two international resolutions [limiting ocean fertilization experiments]," said Kristina M. Gjerde, a senior high seas adviser for the International Union for Conservation of Nature. "Even the placement of iron particles into the ocean, whether for carbon sequestration or fish replenishment, should not take place, unless it is assessed and found to be legitimate scientific research without commercial motivation. This does not appear to even have had the guise of legitimate scientific research." George told the *Guardian* that the two moratoria are a "mythology" and do not apply to his project.[37]

Later, John Disney, president of the Haida Salmon Restoration Corporation, said George hadn't dumped iron sulfate but a "finely ground dirt-like substance with trace amounts of iron." Michaelson (who earlier alluded to geoengineering as the latest Manhattan Project) would have approved:

> . . . geoengineering has moved from the pages of science fiction to respectable scientific and policy journals. One of the most encouraging proposals today focuses on the creation of vast carbon sinks by artificially stimulating phytoplankton growth with iron "fertilizer" in parts of the Earth's oceans. Another proposal suggests creating miniature artificial "Mount Pinatubos" by allowing airplanes to release dust particles into the upper atmosphere, simulating the greenhouse-arresting eruption of Mount Pinatubo in 1991.[38]

---

36 Ibid.

37 Martin Lukacs, "World's biggest geoengineering experiment 'violates' UN rules." *The Guardian*, 15 October 2012.

38 *Stanford Environmental Law Journal*, January 1998.

Whatever the composition, 10,000 square kilometers of blooms became visible from space. UBC biological oceanographer Maite Maldonado said the huge bloom "scares" her: "If you have a massive bloom or growth of this microscopic algae, you might not have enough oxygen in the water column at certain depths," which could create toxic, lifeless waters.[39]

Did George's iron sulfate "experiment" off British Columbia have anything to do with nano-organisms engineered to seek out iron? (See Chapter 8.) Did the iron have anything to do with the 7.7 magnitude earthquake that originated from the island of Haida Gwaii on October 27, 2012?[40]

In 2010, 3 million liters of Corexit 9500A chemical dispersants had been injected into the Gulf of Mexico after the British Petroleum (BP) *Deepwater Horizon* oil catastrophe.[41] While the cover story was oil spill amelioration, I cannot shake the nagging feeling that the anomalies and non sequiturs in both the British Columbia and Gulf of Mexico events point to biological warfare experimentation, this time on a global scale connected with HAARP's electromagnetic capabilities:

> Dr. Susan Shaw, a marine toxicologist, talked about her recent experience with shrimpers who had been working in the Gulf waters. In an interview on CNN, she addressed the situation of a shrimper who had thrown his net into water, causing the water to splash onto his unprotected skin. She reported that he developed a "headache that lasted 3 weeks, heart palpitations, muscle spasms, bleeding from the rectum . . ." and continued, "and that's what this Corexit does, it ruptures red blood cells, causes internal bleeding and liver and kidney damage . . ."[42]

When it comes to inquiring into national security technologies buried in lies, you'll notice that good questions tend to outpace good answers.

EASTLUND AND HAARP

From the beginning, HAARP has been an *international* endeavor of the U.S. military, British BAE Systems, the USSR/Russia, Canada, Japan, Greenland,

39 "Iron fertilization project stirs West Coast controversy," *CBC News*, October 16, 2002. The February 1, 2010 paper "Iron enrichment stimulates toxic diatom production in high-nitrate, low-chlorophyll areas" supports Maldonado's concern.

40 Terry Wilson, "7.7 Earthquake Hits Where the World's Largest Geoengineering Experiment Took Place." Geo-engineering, National News, October 28, 2012.

41 Amanda Mascarelli, "Deepwater Horizon Dispersants Lingered in the Deep." *Scientific American*, January 27, 2011.

42 H.P. Albarelli Jr. and Zoe Martell, "Corexit Tied To 'Dengue Fever' In Florida? Outbreak Leads Back To CIA And Army Experiments." Rense.com, July 22, 2010.

Norway, and Finland.[43] Of course, it all began during the Cold War that set the mold for other "silent" international wars like the War on Drugs and Terrorism.[44]

In *HAARP: The Ultimate Weapon*, Smith goes into considerable detail regarding Russian plasma physicist Roald Zinurovich Sagdeev, director of the East/West Space Science Center at the University of Maryland (one of HAARP's university contractors), where plasma physicist Dennis Papadopoulos[45] has been a professor since 1979. Both scientists were involved with HAARP *before* Eastlund's patent through the USAF Phillips Laboratory and Office of Naval Research (ONR) and its Naval Research Laboratory (NRL).

Besides directing the East/West Space Science Center, Dr. Sagdeev directed the USSR's Space Research Institute (their NASA) from 1973 to 1988, then advised former Soviet Premier Mikhail Gorbachev during the first five years of *perestroika* or, as it is known militarily, the *convergence*—an interesting term. Sagdeev exemplifies this convergence in that he is a member of the Russian Academy of Sciences *and* a member of the American National Academy of Sciences (NAS).

As the Cold War ended and the Gakona, Alaska HAARP installation was underway, Sagdeev was awarded the Leo Szilard Award for Physics in the Public Interest for reversing the arms race and contributing to *glasnost*, though some believe he was awarded for having guided HAARP and plasma physics under a smokescreen of arms control, including organizing (with Evgany P. Velikhov) the Soviet Scientists' Committee for Peace Against the Nuclear Threat, the very committee that in 1986—just before Eastlund's patent—critiqued U.S. Strategic Defense Initiative (SDI) plans. Smith puts it this way: "In Professor Sagdeev, we see a very interesting connection between HAARP, SDI, plasma physics . . . the environment, and arms control."[46]

The Executive Summary of HAARP (Appendix E) was written and circulated to Navy, Air Force, and the Defense Advanced Research Projects Agency

---

43 Begich and Manning, *Angels Don't Play This HAARP*, 1995.

44 "The boundaries between military and civilian targets, between wartime and peacetime conflicts, already beginning to blur during World War II . . . took on an eerie permanence during the Cold War. Military psychological operations experts were only stating what many Americans already felt when they pointed out that peace had lost much of its previous association with security: peace was 'simply a period of less violent war in which nonmilitary means are predominantly used to achieve certain political objectives' [Department of the Army, "Psychological Operations," Department of the Army Field Manual, FM 33-5, January 1962] . . . [The Cold War] put psychological experts to work understanding the style of warfare (guerrilla movements in the Third World) and guiding the new kind of military mission (counterinsurgency) that the postwar decades produced . . . What was the arms race, after all, if not cultural lag come true in the most terrifying of ways?" — Ellen Herman, *The Romance of American Psychology: Political Culture in the Age of Experts* (University of California Press, 1995).

45 Interviewed on "Masters of the Ionosphere," a BBC/A&E TV program, Dr. Papadopoulos was science consultant to APTI. Clifford Carnicom recommends examining Papadopoulos' briefing paper, "Satellite Threat Due to High Altitude Nuclear Detonations" at www.eisenhowerinstitute.org/dotAsset/280369.pdf

46 Smith, *HAARP: The Ultimate Weapon*. Sagdeev was for a time married to Susan Eisenhower, granddaughter of former President Eisenhower.

(DARPA) for coordination on February 7, 1990. A meeting with the Office of the Defense Director of Research & Engineering (DDR&E) and ONR was held five days later to discuss implementation.

On April 30, 1991, another meeting took place at Hanscom Air Force Base from which the 613-page PL/GP Technical Memorandum No. 195 evolved, detailing how HAARP would work in resonance with other ionospheric heaters to achieve "larger effects" and generate ELF signals able to pass through every living being. A "Blackbeard Team" at Los Alamos National Labs (LANL) was designated to control the satellites HAARP required.[47]

MITRE Corporation produced the first draft of the Environmental Impact Statement (EIS) in February 1993. MITRE is a Navy nonprofit founded in 1958 to oversee defense research in concert with the intriguing JASON Group, a secretive scientific organization with top security clearance established in 1960: "For administrative purposes, JASON's activities are run through the MITRE Corporation, a nonprofit corporation in McLean, Virginia, which contracts with the Defense Department."[48] From the initial EIS came the "Electromagnetic Interference Impact of the Proposed Emitters for the High Frequency Active Auroral Research Project (HAARP)" on May 14, 1993. MITRE made sure that the final EIS bypassed Congress and went straight to the Environmental Protection Agency (EPA), after which it was announced in the Federal Register on July 23, 1993.

Originally, the primary contractor for HAARP was ARCO Power Technologies, Inc. (Atlantic Richfield Oil Company) in Monaco, Pennsylvania, later renamed Advanced Power Technologies, Inc. or APTI. Listed in Dun & Bradstreet's "America's Corporate Families," APTI was composed of a president, CEO, and 25 employees. Begich believes APTI was chosen for its proprietary information, namely their consultant Bernard J. Eastlund and his patents. Needless to say, there had been no "competitive procurement process" and APTI had been granted full privileges and exemptions.[49]

From 1966 to 1974, Eastlund had worked for Controlled Thermal Nuclear Research at the Atomic Energy Commission (AEC). Eastlund was always a key researcher for SDI, aptly nicknamed "Star Wars." Behind Eastlund stood the Center for Security Policy, the Star Wars mothership funded by primary players in the military-industrial complex like McDonnell Douglas, Northrop Grumman, TRW, Lockheed Martin, Smith Richardson, Sara Scaife, and Adolph Coors. Since 1996, Eastlund has been CEO of Eastlund Scientific Enterprises Corporation, specializing in weather modification and other high-tech electromagnetic services.

Tracking HAARP patent transfers is a study in the American intelligence-industrial complex. From APTI, the patents first passed to E-Systems, a Dallas

47 Begich and Manning, *Angels Don't Play This HAARP*, 1995.
48 "JASON (advisory group)," Wikipedia.
49 Smith, *HAARP: The Ultimate Weapon*.

defense contractor noted for its many "retired" and current CIA agents. Then they went to Raytheon Corporation, fourth largest American shadowy defense contractor and third largest aerospace company. To give you an idea of Raytheon's size beyond its 72,000 employees and annual gross of $25 billion, Hughes Aircraft Corporation and General Dynamics are both divisions of Raytheon.[50] Perhaps most importantly, Raytheon is the poster child of the National Security Telecommunications Advisory Committee (NSTAC), comprised of 30 industry leaders appointed by the President to oversee domestic telecommunications and "non-lethal" technologies. The NSTAC reports to the Department of Homeland Security. Raytheon stocks soared after 9/11.

Scrape away the shell game of patent transfers and HAARP appears at the very least to be the love child of Raytheon/NSTAC, the U.S. Air Force, and the Electronic Warfare division of British Aerospace Systems (BAES), the North American subsidiary of UK-based BAE Systems, which in the byzantine world of corporate ties includes Lockheed Martin:

> . . .formed in June 2005 as BAE Systems Electronics and Integrated Solutions by the merger of BAE's Information & Electronic Warfare Systems (IEWS) and Information & Electronic Systems Integration (IESI) units. The former was the Lockheed Martin Aerospace Electronic Systems business, acquired by BAE in 2000.[51]

Another telling connection with BAE is Echelon:

> . . . a signals intelligence (SIGINT) collection and analysis network [since the 1960s] operated on behalf of the five signatory states to the UKUSA Security Agreement (Australia, Canada, New Zealand, the United Kingdom, and the United States, referred to by a number of abbreviations, including *AUSCANNZUKUS* and *Five Eyes*). It has also been described as the only software system which controls the download and dissemination of the intercept of commercial satellite trunk communications.[52]

ALASKA AND THE ARCTIC CIRCLE

Once Alaska—so near the Arctic Circle—became a state in 1959, it was doomed to birth Star Wars, beginning with the secretive Single Stage Rocket Technology (SSTR) project, then the Alaska Aerospace Development Corporation at Kodiak and Polaris missile launching pads at Narrow Cape. The other Polaris

---

50 See its Integrated Defense Systems site at www.raytheon.com/businesses/rids/
51 Wikipedia, "BAE Systems Electronics, Intelligence & Support."
52 Wikipedia, "ECHELON." Also see www.fas.org/irp/program/process/echelon.htm

launching site is the island of Kauai in Hawaii, which became a state later in 1959. Designated a First Strike area during the Cold War, Alaskans had already been inundated with early warning radar and irradiation. Project Chariot—the Cannikin nuclear test in 1971—contaminated Amchitka Island and the Bering Sea. Add to that the dark history of Fort Greely Military Reserve and Jarvis Creek, the oil assault on their Arctic National Wildlife Refuge, the radar dome on Shemya Island in the Aleutians, Kodiak's 17 toxic dumps, and finally HAARP.

The human rights of Inupiat and Kasigluk Native Americans on the Arctic slope were not at the top of the U.S. military agenda, given that Alaska was perfectly situated with its near vertical magnetic field lines that could be made to act as a "global shield" of charged particles. Neither the nonthermal effects of the 132 high-frequency transmitters nor the chemicals released into the atmosphere in the name of experimental science (titanium, boron, barium, strontium, lithium, europium, calcium, etc.) were taken into account. In fact, violating the 1978 ENMOD Convention (Environmental Modification) by changing the chemical composition of the atmosphere and transporting plumes of particulates in a plasma-ized atmosphere mattered about as much as the grocery list of ABM treaty violations.

Besides the Arctic Circle, the Gakona site held multiple advantages. ARCO owned the gas reserves in North Alaska, and Diamond A Cattle Company—the largest holding company in the United States—the oil shale deposits in North America but the mineral rights to the Alaska North Slope. Then there was the nearby Poker Flat Research Range necessary for rocket launches.

The first phase of HAARP construction was the Ionospheric Research Instrument (IRI), the high-power radio frequency transmitter that you can readily see in any photograph of the Gakona installation. On June 14, 1994, the Secretary of Defense requested an increase in HAARP funding from $5 million to $75 million purportedly to support HAARP's EPT (earth-penetrating tomography)[53] and HAARP's VHF-UHF 3GHz (*billions*) to create artificial ionospheric mirrors (AIMs) for over-the-horizon radar. The first phase was completed in December 1994, after which the second phase of construction ensued in summer 1997, with HAARP's effective radiation power (ERP) increased up to ten gigawatts.[54]

Supposedly, ARCO approached plasma physicist Bernard Eastlund about finding a use for their North Alaska gas reserves. What this story does not include is the American military's close relationship with oil and its many inquiries into the uses that a $30 million *steerable* RFR (radio frequency radiation) transmitter

---

53 "Counterproliferation — Advanced Development" (Project P539 Counterforce). Also see Sandia National Laboratory's SISAR (Subsurface Imaging Synthetic-Aperture Radar) and NASA's LightSAR, a satellite-based imaging radar system with tomographic mapping capabilities and RF heaters of 400MHz (millions).

54 Remember: *Hertz* is a frequency measurement of cycles per second; *watts* measures electrical power. For example, in North America, 120-volt electricity is delivered at 60 Hz in alternating current.

might be put to, particularly at a perfect latitude (the Arctic Circle) where 644 billion cubic meters of natural gas were waiting to heat up 1.7 gigawatts (billions) of bunched waves and with 28,000°C (50,432°F) raise a square kilometer of the ionosphere so that the jet stream could be directed wherever the DoD wanted. As Begich and Manning put it:

> . . . this [HAARP] invention provides the ability to put unprecedented amounts of power in the Earth's atmosphere at strategic locations and to maintain the power injection level, particularly if random pulsing is employed, in a manner far more precise and better controlled than heretofore accomplished by the prior art . . .[55]

Dan Eden (a.k.a. Gary Vey) in "Weapons of Total Destruction" believes that the Gakona, Alaska array 260 miles south of Poker Flat in the rugged, mountainous Alaska Range may be a "public face" array, while the actual classified HAARP project is located at the flat, spacious, and remote Poker Flat rocket range 30 miles north of Fairbanks.

What lends credence to this possibility is that according to HAARP program manager Dr. James Keeney of Kirtland Air Force Base, the 35-acre HAARP array at 62°23'29.66" N, 145°06'58.47" W—1.7 gigawatts of radiated power in the 2.8–10MHz frequency range—was shut down in early May 2013 for four to six weeks until a new GOCO contractor [government-owned contractor-operated] takes over the facility. The U.S. Air Force is still the government owner and DARPA is still on site.[56]

Has the IRI [ionospheric research instrument], like its 180-antenna array, already become anachronistic? Will the Gakona array be dismantled, bulldozed, or be utilized for yet more research? CEO/owner/senior meteorologist Kevin Martin (theweatherspace.com) and his magnetometer are "doing research" at the aging Gakona facility and publishing it at HaarpStatus.com. Martin says that with new heater facilities taking up HAARP's slack, private companies can now buy time for research.[57]

Or is HAARP merely going deeper black under DARPA's auspices?

The mystique that has grown up around HAARP and its unique steering mechanism at the Gakona site has, like Area 51, distracted the public from the global mushrooming of other ionospheric heaters. Whatever is going on at the Gakona installation, ionospheric heater technology is spreading fast.

---

55 Begich and Manning, *Angels Don't Play This HAARP*, 1995.

56 "HAARP Facility Shuts Down," www.arrl.org [American Radio Relay League], July 15, 2013.

57 "TheWeatherSpace.com Signs With HAARP Facility Under A One Year Contract," www.theweatherspace.com, July 18, 2013.

OTHER IONOSPHERIC HEATERS

HAARP differs from other ionospheric heaters in that it is a phased array. As an illegal, over-the-horizon phased-array RADAR weapon able to track hundreds of objects simultaneously, it is capable of tremendous focusing ability due to its *pulsed sequential firing*. That the ionospheric waveguide oscillates naturally at 8Hz (cycles per second) means it is an excellent harmonic carrier of low-frequency sound (LFS) waves into the alpha range of human brains.[58] Because LFS waves are so long, they are virtually impossible to detect. In other words, HAARP provides LFS-wave field strengths that can simultaneously affect large geographic swaths of the population.

Mainstream science will probably chronicle HAARP as having had an 18-year run with two major accomplishments to its credit: bouncing a 40-meter signal off the Moon (2007) and successfully producing a sustained high-density plasma cloud in the Earth's upper atmosphere (2013). But the real breadth of its "success" lies closer to this list from Begich/Manning and Bertell that I have augmented:

(1) Generate extremely low frequency (ELF) waves for communicating with nuclear submarines;

(2) Earth-penetrating tomography: Penetrate land with ELFs in order to search for hidden tunnels and other sites of military interest;

(3) Generate ionospheric lenses to focus large amounts of HF energy, thus providing a means of triggering ionospheric processes that can be exploited for military purposes;

(4) Electron acceleration for infrared (IR) and other optical emissions that can be used to control radio wave propagation properties;

(5) Generate geomagnetic field-aligned ionization to control the reflection/scattering properties of radio waves;

(6) Use oblique heating to produce effects on radio wave propagation, thus broadening the potential military applications of ionospheric enhancement technology;

(7) Change the chemical structure of the upper and lower atmospheres to alter weather;

(8) Affect human mental functioning;

(9) Impact human health and other biological systems.[59]

---

58 The human brain operates in the range of 0.5–40Hz.

59 Other possibilities might be enhancement and electromagnetic pulse (EMP) disruption of communications, satellites, or city grids; shielding a territory from intercontinental ballistic missiles (ICBMs) or other HAARP technologies; and discriminating between incoming objects (missiles, meteors, debris, etc.).

At 43°04'51.75"N/92°48'26.85"E, near the villages of Jiefang (liberation) and Kan'erjing (underground caverns), Google Earth gives us an aerial shot of what might be China's next generation of HAARP arrays. Nathan Cohen, inventor and patent holder of *fractal arrays* and CEO of Fractal Antenna Systems, Inc. (www.fractenna.com), describes the symmetrical fractal shapes in a satellite photograph:

> . . . two banks of three arrays for two separate bands, and one bank of two arrays for another. You can't tell the operational frequencies from the spacings. The panels are many wavelengths across, but we don't know how many. It is a multiband array antenna farm with flat array panels. If we knew the operational frequencies, I could tell you the gain. Not an Arecibo, but still not too shabby would be my guess. Could be an imaging radar for satellite monitoring.[60]

Fractal antennae are indeed the cutting edge in arrays. Also, check out China's Meridian Project from pole to pole at sincedutch.files.wordpress.com/2013/06/tech-25.pdf, as well as *ViewZone* editor Gary Vey's "The Great Grid of China: A technological wonder of the World" (April 12, 2012)[61]—hundreds of square miles of buried fiber optic cable laid out in fractal patterns in north China.

The future is upon us.

Besides the steering kingpin and incoherent scatter facility at Poker Flat rocket range, other heaters and antenna farms are multiplying in a continually evolving list that global observers are trying to keep track of and decipher. For example, Martin Harris of New Zealand on The CON Trail (August 21, 2013) had his eyes open:

> . . .the US Navy antenna farm at the old RNZAF Weedons base, an area that I am very intrigued about for a number of reasons, mostly relating [to] the Sept 2010 earthquake. For some reason the area seems to attract an inordinate number of accidents, fires and suchlike. It also happens to be where preliminary ionospheric research was carried out prior to the establishment of the 'rocket pad' at Birdlings Flat. Halfway between NZ and Antarctica...conjugation point of HAARP...Project One Hop...a connection I wonder?

Here is a partial list of other ionospheric heaters (transmitters) and observatories (receivers), followed by additional notes on some of the sites:

Millstone Hill Radio Observatory (EISCAT-like facility), Westford, MA
42.5792° N, 71.4383° W

Digisonde Portable Sounder (DPS), University of Massachusetts Lowell's Center for Atmospheric Research, MA

60 Gary Vey, "The Shape of the Future." www.viewzone.com/cgrid/cgrid.html, April 4, 2012.
61 www.viewzone.com/cgrid/cgrid22.html

Manages the frequencies of the 3600kW IRI (ionospheric research instrument) transmitter.
42.6333° N, 71.3167° W

Platteville Atmospheric Observatory (first American HAARP), CO
40.2150° N, 104.8222° W

VOA (Voice of America) (27MW), Delano, CA
35°45′15″ N 119°17′7″ W
Ceased broadcasts in October 2007.

High-Power Auroral Stimulation Observatory (HIPAS), Fairbanks, AK (17MW). 64.8378° N, 147.7164° W
Run by the UCLA Plasma Physics Laboratory; supposedly shut down in 2010.

Chatanika Incoherent Scatter Radar Facility, Poker Flats, AK
65.1200° N, 147.4700° W

Poker Flat Research Range, near Chatanika, AK
65° 7′37.10″ N 147°29′37.61″ W
65° 7′48.34″ N 147°28′14.05″ W
+65° 7′55.61″, -147°27′14.98″

Arecibo Ionospheric Observatory, Puerto Rico (20MW)
18°20′39″ N, 66°45′10″ W
Funded by DARPA, construction began in 1960 at an initial cost of $9.7 million. Administered by U.S. Air Force, managed by Cornell University, then SRI International, Universities Space Research Association, and Metropolitan University (Puerto Rico), under contract to the National Science Foundation (NSF).

São Luiz Space Observatory, Brazil
-2°35′40.47″, -44°12′35.90″
Operational since August 1998. Coherent Back-Scatter Radar of 50MHz (RESCO) installed at the Space Observatory of São Luís / INPE. Network antenna has 768 dipoles to concentrate energy in a narrow beam.

Jicamarca Radio Observatory, Lima, Peru
11°57′6″ S, 76°52′27″ W

SIBNIEE Novosibirsk, Russia
55°0′26.41″ N, 83°1′50.42″ E

SURA, 100 km east of Nizhny Novgorod, Russia
56°7′9.70″ N, 46°2′3.66″ E
56°08′ N, 46°06′ E
Commissioned in 1981. 190MW of radiated power (ERP), shortwave.

ISTP Institute of Solar-Terrestrial Physics
Irkutsk, Russia
51°45′35.99″ N, 102° 6′34.86″ E

SAO RAS Special Astrophysical Observatory of the Russian Academy of Science
Nizhnij Arkhyz, Karachai-Cherkessian Republic, Russia
43°49′44.07″ N, 41°35′3.90″ E

VNITS VEI, Moskovskaja Oblast' Istra, Russia
55°55′26.25″ N, 36°49′8.46″ E

Tesla Tech Arrays (TTAs) Sychëvka, Moskovskaya Oblast, Russia
55.55′26.49″ N, 36.49′11.38″ E

VOLNA GP-120, near Nakhodka, Russia
42°51′42.69″ N, 132°36′50.04″ E

PRAO Pushchino Radio Astronomy Observatory, Moskovskaja Oblast' Pushchino, Russia
54°49′29.02″ N, 37°37′59.06″ E

Duga Radar Array ("Russian Woodpecker"), Chernobyl, Ukraine
51.3896° N, 30.0991° E

UTR-2 URAN (VLBI) IRA Institute of Radio Astronomy NAS Ukraine, Kharkov, Ukraine 49°38′5.53″ N, 36°56′8.11″ E (UTR-2)
49°40′27.47″ N, 36°17′32.24″ E (URAN-1)
49°37′51.17″ N, 34°49′29.80″ E (URAN-2)
51°28′20.66″ N, 23°49′36.92″ E (URAN-3)
46°23′46.11″ N, 30°16′22.52″ E (URAN-4)

URDF-3 (Unidentified Research and Development Facility-3) Baikal-1, Semipalatinsk, Kazakhstan
50.1167° N, 78.7167° E

SuperDARN (Dual Auroral Radar Network), Saskatoon, Saskatchewan, 52.1311° N, 106.6353° W; Kapuskasing, Ontario, 49.4167° N, 82.4333° W; Prince George, Canada, 53.9136° N, 122.7502° W; Goose Bay, Newfoundland, 53.3019° N, 60.4167° W; Stokkseyri, Iceland, 63.8355° N, 21.0505° W; Hankasalmi, Finland, 62.3889° N, 26.4361° E; Kodiak Island 57.4667° N, 153.4333° W and King Salmon, Alaska 58.6883° N, 156.6614° W by the Japanese; Unwin Radar at Awarua, New Zealand, 35° 35' 0" South, 173° 51' 0" East. Radar at Wallops Island, Virginia, 37.93° N, -75.47° E, was the first to be constructed using the Twin-Folded-Dipole-Antenna design.

EISCAT (European Incoherent Scatter Scientific Association, 48MW, 69°35′10.67″ N, 19°13′28.62″ E), Svalbard, Norway

The Ramfjordmoen facility (at Ramfjord, 20 km south of Tromsø, Norway) is the HAARP of Northern Europe. Began operations in 1981. From the EISCAT website: "The Heater is used for ionospheric modification experiments applying high-power transmissions of high-frequency electro-magnetic waves to study plasma parameters in the ionosphere . . . [T]hese high-power electromagnetic waves, which are transmitted into the ionosphere with high-gain antennas, heat the electrons and thus modify the plasma state. To create plasma turbulence, the transmitted frequencies have to be close to the plasma resonances, which are 4 to 8 MHz."[62]

Operates three incoherent scatter radar systems at 224MHz, 931MHz in Northern Scandinavia, and one at 500MHz. Additional receiver stations are located in Sodankylä, Finland, and Kiruna, Sweden. Funded and operated by research institutes and councils of Norway, Sweden, Finland, Japan, China, UK, and Germany.

Tromsø, Norway is where the "Norway Spiral" was seen in December 2009.

ALWIN MST Radar, Andøya, Norway
+69° 17′54.41″, +16° 2′31.48″

ALOMAR Observatory, Andøya, Norway
+69° 16′42.38″, +16° 00′33.36″

HiScat/Teracom, Sweden
55°49 N, 13°44 E

Nerc MST Radar Facility (NMRF), Carmarthenshire, Wales, UK
51.8605° N, 4.3031° W [+52° 25′28.26″, -4° 00′19.59″?]

---

62 HAARP = 2.8–10MHz.

IAP Leibniz Institute of Atmospheric Physics, Kühlungsborn, Germany
54° 7′6.03″ N, 11°46′11.88″ E

TIRA, FGAN-FHR, Fraunhofer Wachtberg, Germany
50°39′1.59″ N, 7° 7′46.87″ E

Jindalee Operational Radar Network (JORN), Laverton, West Australia
28°19′36.29″ S, 122°0′18.84″ E
[Over-The-Horizon Radar (OTH-R) facility]

JORN 2 Long Reach, Australia
-23°39′28.9692″ S, 144°08′43.5552″ E

Leonora, Australia
28°19′02.5608″ S, 122°50′36.4416″ E

Alice Spring, Australia
22°58′03.2196″ S, 134°26′52.5732″ E

Harold E. Holt Naval Communication Station, Exmouth, West Australia
21.9333° S, 114.1333° E

Birdlings Flat, New Zealand
43.8167° S, 172.6833° E
Four radar systems: a meteor radar system, a mesospheric radar system, an ionization radar system and a tropospheric wind profiler.

EQUatorial Ionospheric Study (EQUIS II), Kwajalein Atoll, Marshall Islands
8.7167° N, 167.7333° E
Run by Cornell University and NASA. Very near the equator. Four scientific missions investigate nighttime plasma structures, electrodynamics, and mesospheric scattering processes. Six rockets carry trimethyl aluminum (TMA), a luminous tracer of atmospheric motions that forms milky white clouds in the night sky.[63]

National MST Radar Facility (NMRF), Andra Pradesh, India
13°27′26.68″ N, 79°10′30.74″ E

63 "NASA Successfully Completes Launches for EQUIS II Sounding Rocket Campaign." *NASA News*, press release, September 20, 2004.

SouthEast Asia Low-latitude Ionospheric Network (SEALION) — part of a NICT Ionosonde network monitoring the plasma bubble for IRI

China Research Institute of Radiowave Propagation (CRIRP), Ionospheric Laboratory, Xinjiang (Sinkiang) Region, China
40°24′15.91″ N, 93°38′09.74″ E

Zhong Shan Antarctic Polar Station, China
69°22′44″ S, 76°22′40″ E

Sheshan Radio Observatory (EISCAT-like facility), Shanghai, China
31°05′47″ N, 121°11′16″ E

Ionosphere Observation Network
Chung Li National Central University, Taiwan
Taoyuan County, Jhongli City, Taiwan
24°58′3.73″ N, 121°11′10.59″ E

Mu Shigaraki Obervatory, RISH, Kyoto University, Uji City, Japan
34°51′ N, 136°06′ E

EAR Equatorial Atmosphere Radar, Sumatera Barat, Indonesia
National Institute of Aeronautics and Space
Lembaga Penerbangan dan Antariksa Nasional (LAPAN)
Jakarta, Indonesia
0°12′12.81″ S, 100°19′14.88″ E

Besides a low-frequency band, the *HIPAS Observatory* maintains two other bands. One is a high-frequency band that can turn the Aurora Borealis—the military's "outdoor plasma lab"—into a HF radar emitter with a high-frequency carrier wave in the shortwave band between 2.8–10MHz[64]:

> . . .the auroral electro-dynamic circuit of the electrojet carries toward the earth .1 to 1 million megawatts of power, the equivalent of a hundred to a thousand large power plants. By exploring the properties of the auroral ionosphere as an active, non-linear medium, the primary range of the HF transmitter which is confined to a frequency range of 2.8 to 10MHz can be down-converted in frequency to coherent low-frequency waves spanning five decades and up-converted to infrared and visible photons. As a result the HAARP HF transmitter

64 The National Telecommunications and Information Administration (NTIA), a branch of the Department of Commerce, assigns frequencies federal radio frequencies. It also oversees new and emerging telecommunications technologies and "performs long-term research to explore uses of higher frequency spectrum" (Wikipedia).

> can generate sources for remote sensing and communications spanning sixteen decades in frequency.[65]

"Remote sensing and communications spanning sixteen decades in frequency" must refer to HIPAS' *optical* band, which can create high-altitude plasma holographs in the infrared portion of the spectrum—from artificial auroras like the "airglow" Christmas colors seen in Texas on November 10, 1991[66], to "holographic cities" like those appearing twice on the shores of eastern China, once in 2005 at Penglai City[67] and again in 2011 at Huangshan City[68]. The optical band may also bear upon "optical and advanced infrared (IR) countermeasures systems and technologies,"[69] namely infrared and laser weapons applications.

TOO LITTLE, TOO LATE

Early on, public inquiry into things HAARP and chemtrails was consigned to the outer darkness of the conspiracy fringe. Few heard about ENMOD's reassertion in the UN Framework Convention on Climate Change (UNFCCC) signed at the Earth Summit in Rio in 1992, and fewer still about how geoengineering was a done deal in the UN-backed 1999 Intergovernmental Panel on Climate Change (IPCC) Special Report "Aviation and the Global Environment," then repeated in the November 2011 UNESCO report "Geoengineering the Climate: Research Questions and Policy Implications." (Both report covers were graced by photographs of jumbo jets spewing atmospheric aerosols otherwise known as chemtrails.)

Nor was a peep heard in the American media when the European Parliament on January 28, 1999 called for (1) international conventions over non-lethal weapons, (2) an examination of HAARP by an international independent body, and (3) an international ban *on all developments and deployments of weapons which might enable any form of manipulation of human beings.*

---

65 "HAARP Research and Applications," Technical Information Division of the Naval Research Laboratory, Washington, D.C., June 1995.

66 "The US Navy has also been carrying on High Power Auroral Stimulation (HIPAS) research in Alaska. Through a series of wires and a 15-meter antenna, they have beamed high intensity signals into the upper atmosphere, generating a controlled disturbance in the ionosphere. In November 10, 1991, 'Christmas colors' were seen in the sky above Texas, and scientists admitted that the ionosphere must have been weakened at the time, so that the electrically charged particles hitting the earth's atmosphere created the highly visible light called airglow."

67 "Rare Mirage Lasts for 4 Hours off East China Shore," english.cri.cn/811/2006/05/07/421@85556.htm

68 "China Mirage City: Hologram, Project Blue Beam?" *Beyond the Curtain*, June 22, 2011.

69 See "Optical and Infrared Detection and Countermeasures," *The DTIC [Defense Technical Information Center] Review*, October 1996.

> . . . [the Committee on the Environment, Public Health and Consumer Protection] regards the U.S. military ionospheric manipulation system, HAARP, based in Alaska, which is only a part of the development and deployment of electromagnetic weaponry for both external and internal security use, as an example of the most serious emerging military thread to the global environment and human health, as it seeks to interfere with the highly sensitive and energetic section of the biosphere for military purposes, while all of its consequences are not clear, and calls on the Commission, Council, and Member States to press the U.S. Government, Russia, and any other state involved in such activities to cease them, leading to a global convention against such weaponry . . .[70]

On October 2, 2001—less than a month after 9/11—Representative Dennis Kucinich (D-OH) had the foresight to grasp the peril of this latest military program cloaked as Solar Radiation Management (SRM). Kucinich challenged government, military, and environmental agencies during the 1st Session of the 107th Congress by introducing the Space Preservation Act (HR2977):

> . . . to preserve the cooperative, peaceful uses of space for the benefit of all humankind by permanently prohibiting the basing of weapons in space by the United States, and to require the President to take action and implement a world treaty banning space weapons.

To date, this is the only bill that names chemtrails as an "exotic weapons system" along with particle beams, electromagnetic radiation, plasma, ELF/ULF radiation, and mind control. Not surprisingly, the bill was stalled in committee. On January 23, 2002, during the 2nd Session of the 107th Congress, Rep. Kucinich submitted the revised HR3616 Space Preservation Act of 2002 in which all mention of exotic weapons, including chemtrails, had been removed. Similar to how HAARP was slipped in as a research project, Section 6, "Space-based Nonweapons Activities" slipped "defense activities" under corporate development of space and planet Earth:

> Nothing in this Act may be construed as prohibiting the use of funds for
> (1) space exploration;
> (2) space research and development;
> (3) testing, manufacturing, or production that is not related to space-based weapons or systems; or
> (4) civil, commercial, or defense activities (including communications, navigation, surveillance, reconnaissance, early warning, or remote sensing) that are not related to space-based weapons or systems.

---

70 European Parliament Committee on Foreign Affairs, Security and Defense Policy Report, January 14, 1999.

This bill too died, only to be resurrected during the 108th Congress on December 8, 2003 as HR3657, "Space Preservation Act of 2003"—which also died. On May 18, 2005 with 34 co-sponsors, Kucinich submitted HR2420, "Space Preservation Act of 2005," to the 109th Congress only to watch it too be put to death. He remained in office until unseated by Democrat Marcy Kaptur in the March 2012 primary.

Too little, too late. By the time people began waking up to chemclouds in increasingly white skies in 1998, everything HAARP was protected by unclassified civilian status and embedded media so as to circumvent treaties, public scrutiny, and congressional oversight.

Star Wars and its Death Star had finally arrived.

CHAPTER TWO

Deconstructing Eastlund's 1987 Patent for HAARP

Bernard Eastlund wrote me an e-mail. . . and he said that to heat up the atmosphere with HAARP was very difficult, it would go right through the atmosphere, unless you put some element in that airspace that it could heat and he suggested that polymers would work very well in allowing HAARP to be directed to heat certain sections of the atmosphere. And in fact, we've been seeing . . . cobweb-like material, polymer material all over the United States and other locations, in conjunction with airplanes flying overhead emitting something out of the back end of them. And Eastlund went further and said that heat generation works by adding magnetic iron oxide to the polymer . . .

— William Thomas on *Coast-To-Coast AM* with George Knapp, December 20, 2009

Great support is lent to the hypothesis that Tesla's work and papers were systematically hidden from public view in order to protect the trail of this top-secret work, which today is known as Star Wars.

— Marc Seifer, *Wizard: The Life and Times of Nikola Tesla*, 2001

Solar energy is the positive monopole of the magnetic atom that comes to earth as a spiral; another spiral starts from the earth to the sun with a negative pole to close the cycle. The reflection of this energy, combined with the pulse that characterizes it, creates life. Everything, animate and inanimate, is marked by its own rhythm.

— Pier Ighina (1908–2004)

In December 2012, as Americans awaited the Winter Solstice and end of the world as we know it, scientist Clifford Carnicom and I deconstructed Bernard Eastlund's August 11, 1987 U.S. Patent #4,686,605, "Method and Apparatus For Altering a Region In the Earth's Atmosphere, Ionosphere, and/or Magnetosphere."

Carnicom's background includes a Bachelor of Science *cum laude* degree in Surveying and Photogrammetry from the Civil Engineering Department at California State University, Fresno, California; postgraduate studies at Ohio

State University and Washington University; an Associate of Sciences degree and a Forest Engineering vocational degree from College of the Redwoods, Eureka, California. His education encompasses a wide variety of disciplines, including geodetic science, advanced mathematics, engineering, statistics, physical sciences, computer science, and the life sciences. For 15 years, he worked for the U.S. Department of Defense, Bureau of Land Management, and Forest Service; was elected Employee of the Year and Supervisor of the Year at the Defense Mapping Agency Aerospace Center; and received the Geodetic Sciences Departmental Award for outstanding technical, managerial, and cost-effective performance.

While guiding me through the dense language of Eastlund's patent, Carnicom pointed out patent references to chemtrails and how *electron cyclotronic resonance* is tied to possible "national security" agendas. Readers may follow along by turning to the patent at the end of this chapter. Don't be intimidated by the science; simply follow along the best you can. The scientifically minded will no doubt see further into the patent than I have been able to see. Let's begin with the Abstract (overview):

> *Abstract:*
> A method and apparatus for altering at least one selected region that normally exists above the earth's surface. The region is excited by *electron cyclotron resonance heating* to thereby increase its charged particle density. In one embodiment, circularly polarized electromagnetic radiation is transmitted upward in a direction substantially parallel to and along a field line which extends through the region of plasma to be altered. The radiation is transmitted at a frequency which excites *electron cyclotron resonance* to heat and accelerate the charged particles. This increase in energy can cause ionization of neutral particles which are then absorbed as part of the region, thereby increasing the charged particle density of the region. [Emphasis added.]

The term "embodiment" throughout the patent is a patent term meaning the manner in which an invention can be made, used, practiced or expressed.

The term *plasma* is the fourth state of matter. Stars like our Sun and its solar wind, the space between planets and stars, the intergalactic medium, nebulae, astrophysical jets (luminous ejected plasma), and the ionosphere are all made up of plasma, as are lightning, St. Elmo's Fire, polar aurorae and polar wind.

Plasma is both a gas and a fluid containing charged particles, either positive or negative electrons or ions. *Heat further ionizes it.* Strong electromagnetic fields via laser or microwave generator can also induce ionization. Given that plasma is an electrically neutral medium of positive and negative particles that can be artificially produced if enough heat is applied *and* it can be contained (or steered)—neon signs, plasma TVs, fluorescent and arc lamps, rocket ion thrusters and heat shields, ozone generators, plasma balls, laser-produced plasmas, Tesla coil arcs, inductively coupled plasmas for optical emission spectroscopy, magnetically induced plasmas, etc.—*controlling* it is a highly attractive incentive for the military-industrial complex.

The first section of Eastlund's patent lays out the vast scope of HAARP: to alter regions of the Earth's atmosphere, ionosphere, and magnetosphere, the telling word being *alter*. As Chapter 1 indicated, attempts to alter upper atmosphere regions by "trapping" electrons and ions have been ongoing: raising electron temperatures by hundreds of degrees, establishing artificial belts, etc. It is within this context that Eastlund first mentions *barium*, one of the metals consistently found in precipitation samples after heavy aerosol spraying:

> It has also been proposed to release large clouds of barium in the magnetosphere so that photoionization will increase the cold plasma density, thereby producing electron precipitation through enhanced whistler-mode interactions.

To review the atmospheric regions important to this study, the *troposphere* extends to 6–20 km/3–12 mi., the *ionosphere* begins at 80 km (49 mi.) and continues to 690 km/428 mi., and the *magnetosphere* overlaps the ionosphere and extends into space to 60,000 km/37,282 mi. *toward* the Sun and over 300,000 km/186,411 mi. *away from* the Sun (nightward) as the Earth's magnetotail.

Under column 1, "Description," we learn that electromagnetic radiation will be transmitted from the Earth's surface along "naturally occurring, divergent magnetic field lines extending through regions to be altered" (see Fig. 1–4). These natural magnetic field lines constitute electromagnetic "highways" between the upper and lower atmospheres. The arrows in the Figures indicate that the "highway" is *two-way*: two artificially created "magnetic mirrors" trap[1] electrons and ions so they oscillate *back and forth* around the magnetic field lines in "heliacal paths" having both magnitude and direction.

In column 5, Eastlund details various methods of heating—ohmic magnetic compression, shock waves, magnetic pumping—but the best for HAARP is electron cyclotron resonance. Pass a certain radio-frequency current through a concentric coil with the axial magnetic field confining the plasma and in each half-cycle of rotation about the magnetic field lines, the charged particles acquire energy from the oscillating electric field associated with the frequency.

Earth's atmosphere—upper and lower—is to be transmuted into a "fully ionized plasma" by means of *heat*, in particular *electron cyclotron resonance*.

In his 1990 book *Cross Currents: The Perils of Electropollution, The Promise of*

---

1 Eastlund: "The term 'trapped' herein refers to situations where the force of gravity on the trapped particles is balanced by magnetic forces rather than hydrostatic or collisional forces." Envision neon gas in a tube.

Electromedicine, Dr. Robert O. Becker discusses cyclotron resonance, specifically in the section "Enter the Resonance Concept" (pages 234–239):

> Cyclotron resonance can be explained as follows, albeit in a somewhat simplistic fashion: If a charged particle or ion is exposed to a steady magnetic field in space, it will begin to go into a circular, or orbital, motion at right angles to the applied magnetic field. The speed with which it orbits will be determined by the ratio between the charge and the mass of the particle and by the strength of the magnetic field . . .
>
> Cyclotron resonance may be produced any time there is a steady magnetic field combined with an oscillating electric or magnetic field acting on a charged particle . . .
>
> Cyclotron resonance is a mechanism of action that enables very low-strength electromagnetic fields, acting in concert with the Earth's geomagnetic field, to produce major biological effects by *concentrating the energy in the applied field upon specific particles*, such as the biologically important ions of sodium, calcium, potassium, and lithium. [Becker's emphasis]
>
> Electron cyclotron resonance heating has been used in experiments on the earth's surface to produce and accelerate plasmas in a diverging magnetic field. [NASA's H.G.] Kosmahl et al. showed that power was transferred from the electromagnetic waves and that a fully ionized plasma was accelerated with a divergence angle of roughly 13 degrees . . . see "Plasma Acceleration with Microwaves Near Cyclotron Resonance," Kosmahl et al. *Journal of Applied Physics*, Vol. 38 No. 12, Nov. 1967, pp. 4576–4582.

The strength of the Earth's geomagnetic field is actually very low (0.2–0.6 gauss[2]), and the strength of our *biofield* even lower. This means that all living creatures are incredibly vulnerable to the electropollution of powerlines, cell towers, and other frequencies, and that's not even counting HAARP's electron cyclotronic resonance heating capability that operates in the lower frequencies:

> . . . the frequencies for the oscillating fields that are needed to produce resonance with the biologically important ions turn out to be *in the ELF region*.
>
> The ELF frequencies—0–100 Hz—become the most significant part of our electromagnetic environment. The apparent ability of the body to demodulate all higher frequencies, including microwaves, substantiates this. Cyclotron resonance provides an understandable and valid mechanism of action for the biological effects of both normal and abnormal electromagnetic fields.

What people don't usually realize is that long-wave ELF transmissions like those used by HAARP have a peculiar, *concerning* property, which in his

2 A gauss (G) measures the magnetic flux density of a magnetic field.

not-to-be-missed 1985 book *The Body Electric*, Becker describes as:

> Because of their interaction with the ionosphere, even weak signals in this frequency range (from 0.1 to 100 cycles per second) travel all the way around the world without dying out. If an innate frequency selector is operating within this band, reception should be the same anywhere on earth.[3]

Following Becker's indications that major biological effects can be produced by "concentrating the energy in the applied field upon specific particles, such as the biologically important ions of sodium, calcium, potassium, and lithium," Carnicom has examined cyclotronic resonance as it relates to the relationship between an atmospheric ELF harmonic he has detected and the potassium ion:

> . . . the fifth harmonic of the ELF that has been repeatedly measured over a period of several years corresponds to the cyclotronic resonant frequency of potassium. This fifth harmonic, along with numerous other harmonics, is a regular component of the ELF radiation that is under measurement at this time.

Carnicom's assertion that the potassium ion is being targeted for "biological interference within people over large regions of the earth's surface" is serious. He follows it up with spectral analysis and equations, after which he drops the bombshell:

> It can also be expected that variations in the magnetic field of the earth can lead to other potential resonance conditions in various regions or latitudes. It is therefore not unexpected to find large regional health issues that will correlate with variations in the magnetic field strength of the earth. Certain ions are expected to be disrupted in some areas of the globe more than others.

In short, HAARP and other ionospheric heaters around the globe appear to be infusing multiples of artificial 4Hz ELF propagation into our modified atmosphere to create a cyclotronic resonance that can affect biologically important ions of sodium, calcium, potassium and lithium.

Remember: When it comes to living beings, *the lower the frequency, the higher the power.*

With this biological aspect of HAARP's cyclotronic resonance capability in mind, we return to Eastlund's patent.

In Column 3, "Disclosure of the Invention," Eastlund explains that to increase plasma density, electrons must either be heated or artificially created in the upper atmosphere. Circularly polarized electromagnetic radiation is

3 Robert O. Becker, MD, and Gary Selden. *The Body Electric: Electromagnetism and the foundation of Life*. Harper, 1985.

first transmitted up from "where a naturally occurring dipole magnetic field (force) line intersects the earth's surface"—as from the HAARP installation on the Great Circle. By transmitting at a frequency based on the "gyrofrequency of the charged particles" (right-hand circular polarization for the Northern Hemisphere, left-hand for the Southern), electron cyclotron resonance will cause charged particles to heat up and accelerate as they corkscrew up their heliacal paths along magnetic field lines.

Heat, acceleration, and density wouldn't have anything to do with "global warming," would they?

But to make sure the density of charged particles is sufficient, Eastlund now refers more directly to what we know as the chemtrail delivery system in which chemtrails dump and HAARP pumps:

> Sufficient energy is employed to cause ionization of neutral particles (molecules of oxygen, nitrogen and the like particulates, etc.) which then become part of the region thereby increasing the charged particle density of the region. This effect can further be *enhanced by providing artificial particles, e.g. electrons, ions, etc., directly into the region to be affected from a rocket, satellite, or the like to supplement the particles in the naturally occurring plasma.* [Emphasis added.]

Ion-charged particle density. Charge electron density and you can then deploy different frequencies for different agendas.

Once the charged particle density becomes a plasma layer, the main beam of polarized electromagnetic radiation continues to corkscrew and transmit to the ionosphere as a *second* electromagnetic ground- or satellite-based beam locks on from a different source at a different frequency, thus "modulating" the original beam. Eastlund discusses beams (beyond those of the array at the Gakona facility) transmitting up to the "embodiment" to create a resonance and propagate yet another plasma wave or "embodiment," thus modulating various new waves in a "highly nonlinear fashion."

> The amplitude of the frequency of the main beam and/or the second beam or beams is modulated in resonance with at least one known oscillation mode to propagate a known frequency wave or waves throughout the ionosphere.

Trap electrons and ions, enhance them with artificial particles distributed by aerosols, heat the plasma, and you have an altered atmosphere: "The production of enhanced ionization will also alter the distribution of atomic and molecular constituents of the atmosphere, most notably through increased atomic nitrogen concentration" (column 4, "Best Modes for Carrying Out the Invention").

Fig. 1 shows how power in the upper atmosphere can be generated for operations below. We see the Earth (10), a dipole magnetic field line (11), and geographic locations 13 and 14 to be "chosen based on a particular operation to be carried out," given that the "electron cyclotron resonance heating effect can be made to act on electrons anywhere above the surface of the earth":

> . . . electrons may be already present in the atmosphere, ionosphere, and/or magnetosphere of the earth, or can be artificially generated by a variety of means such as X-ray beams, charged particle beams, lasers, the plasma sheath surrounding an object such as a missile or meteor, and the like. Further, artificial particles, e.g. electrons, ions, etc., can be injected directly into region R from an earth-launched rocket or orbiting satellite . . .

The study of magnetic fields and how they interact with conducting fluids like plasma is called *magnetohydrodynamics (MHD)*[4] and is essential to astrophysics and atmospheric science. Huge electric forces can be created with giant Tesla coil transmitters, after which HF power can be steered by phased array transmitters like HAARP's Ionospheric Research Instrument (IRI). MHD generators (see Fig. 5) can transform thermal energy and kinetic energy directly into electricity. Coupled with superconducting magnets[5], they are capable of developing intense magnetic fields. Column 13:

> A particular advantage for MHD generators is that they can be made to generate large amounts of power with a small-volume, lightweight device. For example, a 1000-megawatt MHD generator can be constructed using superconducting magnets to weigh roughly 42,000 pounds and can be readily air lifted.

The Great Circle in Alaska fits the bill perfectly. As Eastlund describes, a "combination of natural gas and magnetohydrodynamic gas turbine, fuel cell and/or EGD [Electric Generators Direct] electric generators at the point where the useful field lines intersect the earth's surface" is particularly advantageous. However, he also stresses that neither HAARP's cyclotronic resonance of particular atoms nor plasma frequency is "limited to locations where the fuel source naturally exists or where desirable field lines naturally intersect the earth's surface." HAARP can be "dialed in" at any frequency anywhere.

Once generated and maximized, the plasma can be focused in an electromagnetic beam (X-ray, charged particle, laser) "to modify or disrupt microwave transmissions of satellites . . . enhance, interfere with, or otherwise modify communications transmissions . . . over a very large portion of the earth . . . in time periods of two minutes or less." Add to this the possibility

4 Clifford E. Carnicom, "Magnetohydrodynamic (MHD) Considerations," March 12, 2001.

5 The SQUID (superconducting quantum interference device) is an example of a very sensitive magnetometer.

that superconducting magnets and MHD generators working together can be weaponized to create earthquakes:

> The results of this study shows that the action of high-energy electromagnetic pulses [charged particle beams] radiated by MHD generators causes substantial changes in the seismicity of earthquake source zones by ACCELERATING the release of energy stored in the crust due to the activity of natural tectonic processes . . . The data were obtained by MHD firing runs performed in the regions of Garm and Bishkek in Central Asia, and the earthquake catalogues analyzed (in different windows of earthquake magnitude, focal depth and hypocentral distance) correspond to known seismogenic regions, namely southern Tien Shan and Tadjik Depression (1976–1978) and northern Tien Shan and Chu Valley (1983–1989) . . . Particular attention has been given to the search for *possible links between MHD-induced changes in the seismic regime and variations of the natural electromagnetic field* . . .[6] [Emphasis added.]

Given HAARP's "ability to employ and transmit over very wide areas of the earth a plurality of electromagnetic waves of varying frequencies and to change same at will in a random manner," we begin to see what a remarkable weapon this technology is, and how far beyond a civilian aurora "research" study it is. In Column 11, Eastlund claims:

> Thus, this invention provides the ability to put unprecedented amounts of power in the earth's atmosphere at strategic locations and to maintain the power injection level, *particularly if random pulsing is employed*, in a manner far more precise and better controlled than heretofore accomplished by the prior art, particularly by the detonation of nuclear devices of various yields at various altitudes. [Emphasis added.]

Most people do not generally think of communications as a weapon. The military/intelligence concept of C4 [command, control, communications, computers], however, is certainly about war strategy, whether in modern war theatres or domestic operations; we have only to think of surveillance and "non-lethal" targeting of individuals and populations[7]. Also in Column 11, Eastlund speaks to the military mind about HAARP's communication capabilities:

> Further, by knowing the frequencies of the various electromagnetic beams employed in the practice of this invention, it is possible not only to interfere with third party communications but to take advantage of one or more such

6 "Effects of MHD-Generated Electromagnetic Discharges on the Seismic Regime (EM-Quake)." Institute for High Energy Densities (IHED), the United Institute of Physics of the Earth, Russian Academy of Sciences (UIPE) *Periodic Report*, May 2000–May 2002.

7 As was listed in Chapter 1 as items 8 and 9 in the list of HAARP accomplishments.

> beams to carry out a communications network even though the rest of the world's communications are disrupted. Put another way, what is used to disrupt another's communications can be employed by one knowledgeable of this invention as a communications network at the same time. In addition, once one's own communication network is established, the far-reaching extent of the effects of this invention could be employed to pick up communication signals of other [sic] for intelligence purposes. Thus, it can be seen that the disrupting effects achievable by this invention can be employed to benefit by the party who is practicing this invention since knowledge of the various electromagnetic waves being employed and how they will vary in frequency and magnitude can be used to an advantage for positive communication and eavesdropping purposes at the same time.

Eastlund refers to "mirrors," "mirror points," and "mirror force" 19 times throughout the patent. These are references to the artificial ionospheric mirrors (AIMs) made from *plasma balls*. AIMs are pivotal to HAARP over-the-horizon radar, steering, and targeting operations:

> The trapped electrons and ions are confined along the field lines between two magnetic mirrors which exist at spaced apart points along those field lines. The trapped electrons and ions move in helical paths around their particular field lines and "bounce" back and forth between the magnetic mirrors . . .
>
> The charged electrons and ions in the ionosphere also follow helical paths around magnetic field lines within the ionosphere but are not trapped between mirrors . . .
>
> A ring of hot electrons is formed at the earth's surface in the magnetic mirror by a combination of electron cyclotron resonance and stochastic heating . . .
>
> In another embodiment of the invention, electron cyclotron resonance heating is carried out in the selected region or regions at sufficient power levels to allow a plasma present in the region to generate a *mirror force* which forces the charged electrons of the altered plasma upward along the force line to an altitude which is higher than the original altitude. In this case the relevant mirror points are at the base of the altered region or regions . . . [T]he altered plasma can be trapped on the field line between mirror points and will oscillate in space for prolonged periods of time. By this embodiment, a plume of altered plasma can be established at selected locations for communication modification *or other purposes*. [Emphasis added.]

On February 25, 2013—29 years after Eastlund's patent saw the light of day—the U.S. Naval Research Laboratory (NRL) went public with how they had produced a high-density *plasma ball* artificial mirror "to be used for reflection of HF radar and communications signals." The story is that HAARP's 3.6MW high-frequency (HF) transmissions had sustained an artificial mirror for over an hour.

> ... The third harmonic artificial glow plasma clouds were obtained with HAARP using transmissions at 4.34 megahertz (MHz) ... Past attempts to produce electron density enhancements have yielded densities of 4 X 10^5 electrons per cubic centimeter (cm^3) using HF radio transmissions near the second, third, and fourth harmonics of the electron cyclotron frequency. This frequency near 1.44 MHz is the rate that electrons gyrate around the Earth's magnetic field ... The NRL group succeeded in producing artificial plasma clouds with densities exceeding 9 x 105 electrons cm3 using HAARP transmission at the sixth harmonic of the electron cyclotron frequency.[8]

Whether or not this achievement occurred when the NRL said it did or lasted longer than an hour, sustaining AIMs is essential for HAARP's global-scale operations.

As Eastlund draws his history-making patent to a close, he praises the "generation of electricity by motion of a conducting fluid [plasma] through a magnetic field" and the "phenomenal variety of possible ramifications and potential future developments" that will follow, including but not limited to:

- lifting large regions of the atmosphere to an "unexpectedly high altitude," causing missiles to encounter unexpected drag forces and self-destruct;
- assuring global weather modification by altering upper atmosphere wind patterns or solar absorption patterns through "constructing one or more plumes of atmospheric particles which will act as a lens or focusing device";
- "positive environmental effects": after *changing the molecular composition of an atmospheric region,* particular molecules "can be chosen for increased presence"; "small micron-sized particles ... with desired characteristics such as tackiness, reflectivity, absorptivity, etc., can be transported for specific purposes or effects";
- ". . . the earth's magnetic field could be significantly altered in a controlled manner by plasma beta effects ... [or] decreased or disrupted at appropriate altitudes to modify or eliminate the magnetic field in high Compton electron generation (e.g., from high-altitude nuclear bursts) regions."
- plumes to simulate and/or perform a detonation similar to that of nuclear devices.

8 "NRL Scientists Produce Densest Artificial Ionospheric Plasma Clouds Using HAARP," NRL News Release, February 25, 2013.

Below is a complete list of the HAARP patents necessary to this multipurpose weapon system that many consider to be a next-generation Tesla scalar electromagnetic weapon once known as the Tesla howitzer or Tesla Magnifying Transmitter (TMT)—a *longitudinal wave interferometer.* Hopefully, this deconstruction of key points in patent #4,686,605 has helped readers to crawl inside at least some of the patent's language. Now, let's move on to chemtrails and the various HAARP-chemtrail pump-and-dump operations that are affecting us all.

HAARP Patents

U.S. Patent 4686605: Method And Apparatus For Altering A Region In The Earth's Atmosphere, Ionosphere, And/Or Magnetosphere. *Issued: Aug. 11, 1987 Filed: Jan. 10, 1985*

U.S. Patent 5038664: Method For Producing A Shell Of Relativistic Particles At An Altitude Above The Earth's Surface. *Issued: Aug. 13, 1991 Filed: Jan. 10, 1985*

U.S. Patent 4712155: Method And Apparatus For Creating An Artificial Electron Cyclotron Heating Region Of Plasma. *Issued: Dec. 8, 1987 Filed: Jan. 28, 1985*

U.S. Patent 5068669: Power Beaming System. *Issued: Nov. 26, 1991 Filed: Sep. 1, 1988*

U.S. Patent 5218374: Power Beaming System With Printer Circuit Radiating Elements Having Resonating Cavities. *Issued: June 8, 1993 Filed: Oct. 10, 1989*

U.S. Patent 5293176: Folded Cross Grid Dipole Antenna Element. *Issued: Mar. 8, 1994 Filed: Nov. 18, 1991*

U.S. Patent 5202689: Lightweight Focusing Reflector For Space. *Issued: Apr. 13, 1993 Filed: Aug. 23, 1991*

U.S. Patent 5041834: Artificial Ionospheric Mirror Composed Of A Plasma Layer Which Can Be Tilted. *Issued: Aug. 20, 1991 Filed: May 17, 1990*

U.S. Patent 4999637: Creation Of Artificial Ionization Clouds Above The Earth. *Issued: Mar. 12, 1991 Filed: May 14, 1987*

U.S. Patent 4954709: High Resolution Directional Gamma Ray Detector. *Issued: Sep. 4, 1990 Filed: Aug. 16, 1989*

U.S. Patent 4817495: Defense System For Discriminating Between Objects In Space. *Issued: Apr. 4, 1989 Filed: Jul. 7, 1986*

U.S. Patent 4873928: Nuclear-Sized Explosions Without Radiation. *Issued: Oct. 17, 1989 Filed: June 15, 1987*

United States Patent [19]

Eastlund

[11] **Patent Number: 4,686,605**

[45] **Date of Patent: Aug. 11, 1987**

[54] **METHOD AND APPARATUS FOR ALTERING A REGION IN THE EARTH'S ATMOSPHERE, IONOSPHERE, AND/OR MAGNETOSPHERE**

[75] Inventor: **Bernard J. Eastlund,** Spring, Tex.

[73] Assignee: **APTI, Inc.,** Los Angeles, Calif.

[21] Appl. No.: **690,333**

[22] Filed: **Jan. 10, 1985**

[51] **Int. Cl.**[4] **H05B 6/64;** H05C 3/00; H05H 1/46

[52] **U.S. Cl.** .... **361/231;** 89/1.11; 380/59; 244/158 R

[58] **Field of Search** .... 361/230, 231; 244/158 R; 376/100; 89/1.11; 380/59

[56] **References Cited**

PUBLICATIONS

Liberty Magazine, (2/35) p. 7 N. Tesla.

New York Times (9/22/40) Section 2, p. 7 W. L. Laurence.

New York Times (12/8/15) p. 8 Col. 3.

Primary Examiner—Salvatore Cangialosi

Attorney, Agent, or Firm—Roderick W. MacDonald

[57] **ABSTRACT**

A method and apparatus for altering at least one selected region which normally exists above the earth's surface. The region is excited by electron cyclotron resonance heating to thereby increase its charged particle density. In one embodiment, circularly polarized electromagnetic radiation is transmitted upward in a direction substantially parallel to and along a field line which extends through the region of plasma to be altered. The radiation is transmitted at a frequency which excites electron cyclotron resonance to heat and accelerate the charged particles. This increase in energy can cause ionization of neutral particles which are then absorbed as part of the region thereby increasing the charged particle density of the region.

15 Claims, 5 Drawing Figures

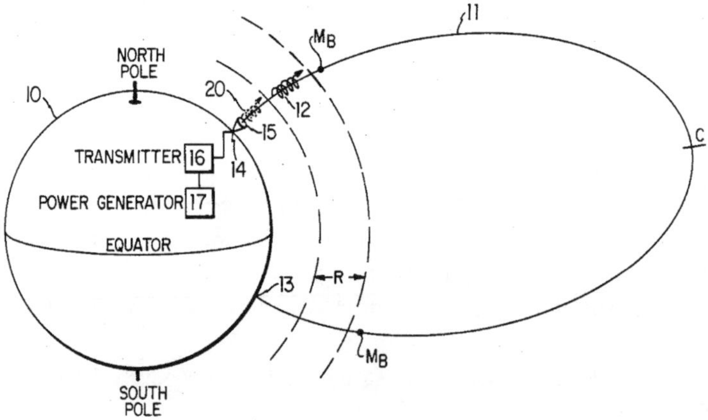

Bernard J. Eastlund Patent #4,686,605

www.scribd.com/doc/27704690/US-Patent-No-4-686-605-HAARP

FIG. 1

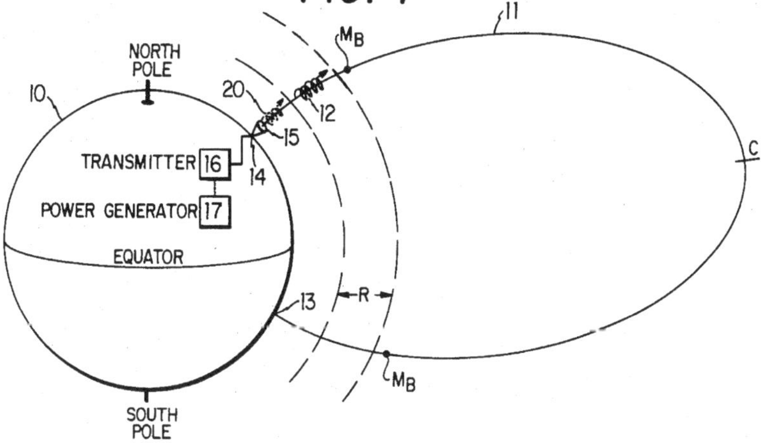

NORTH POLE
10
20
M_B
11
12
15
14
TRANSMITTER
16
POWER GENERATOR
17
EQUATOR
13
R
C
M_B
SOUTH POLE

FIG. 2

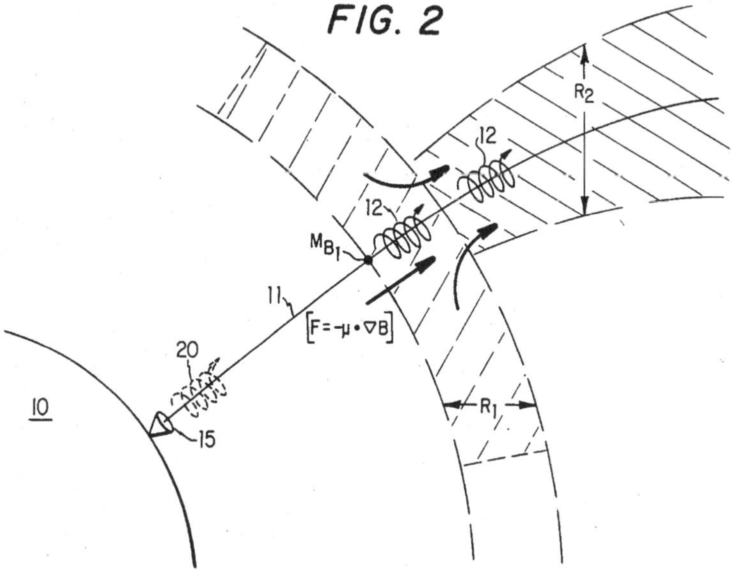

R_2
12
12
M_{B_1}
11
[F=-μ•∇B]
20
10
15
R_1

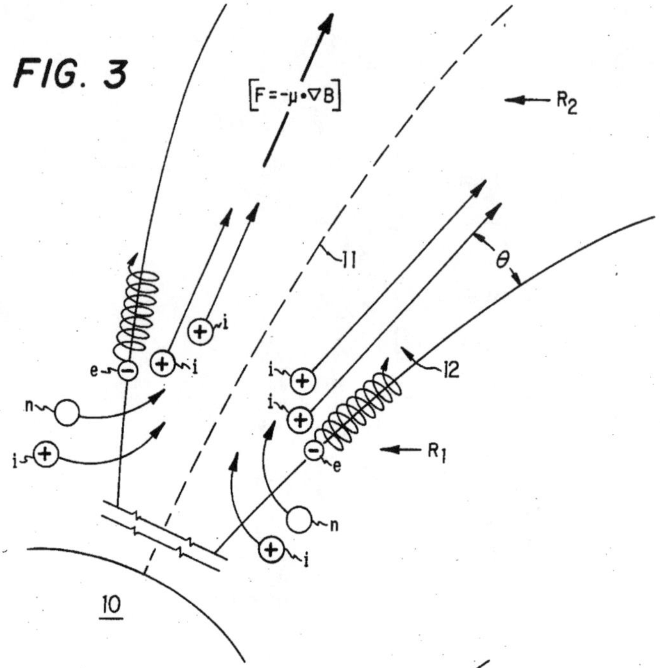

FIG. 3

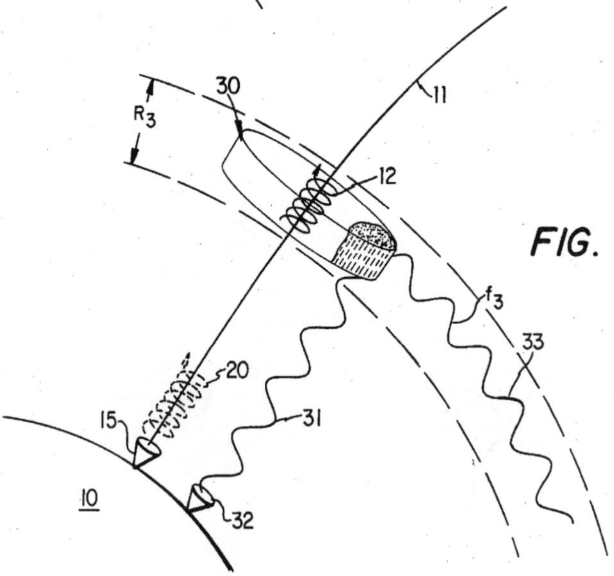

FIG. 4

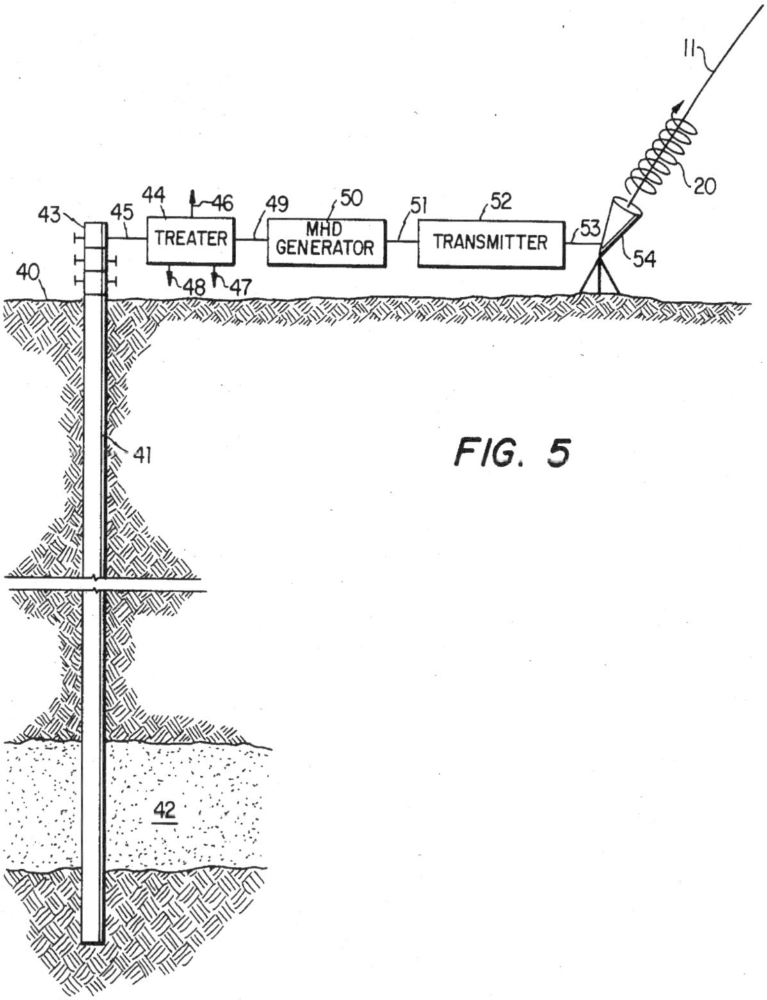
11
20
43
45
44
46
49
50
51
52
53
54
40
48
47
TREATER
MHD GENERATOR
TRANSMITTER
41
42

FIG. 5

METHOD AND APPARATUS FOR ALTERING A REGION IN THE EARTH'S ATMOSPHERE, IONOSPHERE, AND/OR MAGNETOSPHERE

DESCRIPTION

1. Technical Field

This invention relates to a method and apparatus for altering at least one selected region normally existing above the earth's surface and more particularly relates to a method and apparatus for altering said at least one region by initially transmitting electromagnetic radiation from the earth's surface essentially parallel to and along naturally-occurring, divergent magnetic field lines which extend from the earth's surface through the region or regions to be altered.

2. Background Art

In the late 1950's, it was discovered that naturally-occuring belts exist at high altitudes above the earth's surface, and it is now established that these belts result from charged electrons and ions becoming trapped along the magnetic lines of force (field lines) of the earth's essentially dipole magnetic field. The trapped electrons and ions are confined along the field lines between two magnetic mirrors which exist at spaced apart points along those field lines. The trapped electrons and ions move in helical paths around their particular field lines and "bounce" back and forth between the magnetic mirrors. These trapped electrons and ions can oscillate along the field lines for long periods of time.

In the past several years, substantial effort has been made to understand and explain the phenomena involved in belts of trapped electrons and ions, and to explore possible ways to control and use these phenomena for beneficial purposes. For example, in the late 1950's and early 1960's both the United States and U.S.S.R. detonated a series of nuclear devices of various yields to generate large numbers of charged particles at various altitudes, e.g., 200 kilometers (km) or greater. This was done in order to establish and study artifical belts of trapped electrons and ions. These experiments established that at least some of the extraneous electrons and ions from the detonated devices did become trapped along field lines in the earth's magnetosphere to form artificial belts which were stable for prolonged periods of time. For a discussion of these experiments see "The Radiation Belt and Magnetosphere", W. N. Hess, Blaisdell Publishing Co., 1968, pps. 155 et sec.

Other proposals which have been advanced for altering existing belts of trapped electrons and ions and/or establishing similar artificial belts include injecting charged particles from a satellite carrying a payload of radioactive beta-decay material or alpha emitters; and injecting charged particles from a satellite-borne electron accelerator. Still another approach is described in U.S. Pat. No. 4,042,196 wherein a low energy ionized gas, e.g., hydrogen, is released from a synchronous orbiting satellite near the apex of a radiation belt which is naturally-occurring in the earth's magnetosphere to produce a substantial increase in energetic particle precipitation and, under certain conditions, produce a limit in the number of particles that can be stably trapped. This precipitation effect arises from an enhancement of the whistler-mode and ion-cyclotron mode interactions that result from the ionized gas or "cold plasma" injection.

It has also been proposed to release large clouds of barium in the magnetosphere so that photoionization will increase the cold plasma density, thereby producing electron precipitation through enhanced whistler-mode interactions.

However, in all of the above-mentioned approaches, the mechanisms involved in triggering the change in the trapped particle phenomena must be actually positioned within the affected zone, e.g., the magnetosphere, before they can be actuated to effect the desired change.

The earth's ionosphere is not considered to be a "trapped" belt since there are few trapped particles therein. The term "trapped" herein refers to situations where the force of gravity on the trapped particles is balanced by magnetic forces rather than hydrostatic or collisional forces. The charged electrons and ions in the ionosphere also follow helical paths around magnetic field lines within the ionosphere but are not trapped between mirrors, as in the case of the trapped belts in the magnetosphere, since the gravitational force on the particles is balanced by collisional or hydrostatic forces.

In recent years, a number of experiments have actually been carried out to modify the ionosphere in some controlled manner to investigate the possibility of a beneficial result. For detailed discussions of these operations see the following papers: (1) Ionospheric Modification Theory; G. Meltz and F. W. Perkins; (2) The Platteville High Power Facility; Carrol et al.; (3) Arecibo Heating Experiments; W. E. Gordon and H. C. Carlson, Jr.; and (4) Ionospheric Heating by Powerful Radio Waves; Meltz et al., all published in Radio Science, Vol. 9, No. 11, November, 1974, at pages 885–888; 889–894; 1041–1047; and 1049–1063, respectively, all of which are incorporated herein by reference. In such experiments, certain regions of the ionosphere are heated to change the electron density and temperature within these regions. This is accomplished by transmitting from earth-based antennae high frequency electromagnetic radiation at a substantial angle to, not parallel to, the ionosphere's magnetic field to heat the ionospheric particles primarily by ohmic heating. The electron temperature of the ionosphere has been raised by hundreds of degrees in these experiments, and electrons with several electron volts of energy have been produced in numbers sufficient to enhance airglow. Electron concentrations have been reduced by a few percent, due to expansion of the plasma as a result of increased temperature.

In the Elmo Bumpy Torus (EBT), a controlled fusion device at the Oak Ridge National Laboratory, all heating is provided by microwaves at the electron cyclotron resonance interaction. A ring of hot electrons is formed at the earth's surface in the magnetic mirror by a combination of electron cyclotron resonance and stochastic heating. In the EBT, the ring electrons are produced with an average "temperature" of 250 kilo electron volts or kev (2.5×10^9K) and a plasma beta between 0.1 and 0.4; see, "A Theoretical Study of Electron—Cyclotron Absorption in Elmo Bumpy Torus", Batchelor and Goldfinger, Nuclear Fusion, Vol. 20, No. 4 (1980) pps. 403–418.

Electron cyclotron resonance heating has been used in experiments on the earth's surface to produce and accelerate plasmas in a diverging magnetic field. Kosmahl et al. showed that power was transferred from the electromagnetic waves and that a fully ionized plasma was accelerated with a divergence angle of roughly 13 degrees. Optimum neutral gas density was 1.7×10^{14}per cubic centimeter; see, "Plasma Acceleration with Microwaves Near Cyclotron Resonance", Kosmahl et al., Journal of Applied Physics, Vol. 38, No. 12, Nov., 1967, pps. 4576–4582.

DISCLOSURE OF THE INVENTION

The present invention provides a method and apparatus for altering at least one selected region which normally exists above the earth's surface. The region is excited by electron cyclotron resonance heating of electrons which are already present and/or artifically created in the region to thereby increase the charged particle energy and ultimately the density of the region.

In one embodiment this is done by transmitting circularly polarized electromagnetic radiation from the earth's surface at or near the location where a naturally-occurring dipole magnetic field (force) line intersects the earth's surface. Right hand circular polarization is used in the northern hemisphere and left hand circular polarization is used in the southern hemisphere. The radiation is deliberately transmitted at the outset in a direction substantially parallel to and along a field line which extends upwardly through the region to be altered. The radiation is transmitted at a frequency which is based on the gyrofrequency of the charged particles and which, when applied to the at least one region, excites electron cyclotron resonance within the region or regions to heat and accelerate the charged particles in their respective helical paths around and along the field line. Sufficient energy is employed to cause ionization of neutral particles (molecules of oxygen, nitrogen and the like, particulates, etc.) which then become a part of the region thereby increasing the charged particle density of the region. This effect can further be enhanced by providing artificial particles, e.g., electrons, ions, etc., directly into the region to be affected from a rocket, satellite, or the like to supplement the particles in the naturally-occurring plasma. These artificial particles are also ionized by the transmitted electromagnetic radiation thereby increasing charged particle density of the resulting plasma in the region.

In another embodiment of the invention, electron cyclotron resonance heating is carried out in the selected region or regions at sufficient power levels to allow a plasma present in the region to generate a mirror force which forces the charged electrons of the altered plasma upward along the force line to an altitude which is higher than the original altitude. In this case the relevant mirror points are at the base of the altered region or regions. The charged electrons drag ions with them as well as other particles that may be present. Sufficient power, e.g., 10^{15}joules, can be applied so that the altered plasma can be trapped on the field line between mirror points and will oscillate in space for prolonged periods of time. By this embodiment, a plume of altered plasma can be established at selected locations for communication modification or other purposes.

In another embodiment, this invention is used to alter at least one selected region of plasma in the ionosphere to establish a defined layer of plasma having an increased charged particle density. Once this layer is established, and while maintaining the transmission of the main beam of circularly polarized electromagnetic radiation, the main beam is modulated and/or at least one second different, modulated electromagnetic radiation beam is transmitted from at least one separate source at a different frequency which will be absorbed in the plasma layer. The amplitude of the frequency of the main beam and/or the second beam or beams is modulated in resonance with at least one known oscillation mode in the selected region or regions to excite the known oscillation mode to propagate a known frequency wave or waves throughout the ionosphere.

BRIEF DESCRIPTION OF THE DRAWINGS

The actual construction, operation, and apparent advantages of this invention will be better understood by referring to the drawings in which like numerals identify like parts and in which:

FIG. 1 is a simplified schematical view of the earth (not to scale) with a magnetic field (force) line along which the present invention is carried out;

FIG. 2 is one embodiment within the present invention in which a selected region of plasma is raised to a higher altitude;

FIG. 3 is a simplified, idealized representation of a physical phenomenon involved in the present invention; and

FIG. 4 is a schematic view of another embodiment within the present invention.

FIG. 5 is a schematic view of an apparatus embodiment within this invention.

BEST MODES FOR CARRYING OUT THE INVENTION

The earth's magnetic field is somewhat analogous to a dipole bar magnet. As such, the earth's magnetic field contains numerous divergent field or force lines, each line intersecting the earth's surface at points on opposite sides of the Equator. The field lines which intersect the earth's surface near the poles have apexes which lie at the furthest points in the earth's magnetosphere while those closest to the Equator have apexes which reach only the lower portion of the magnetosphere.

At various altitudes above the earth's surface, e.g., in both the ionosphere and the magnetosphere, plasma is naturally present along these field lines. This plasma consists of equal numbers of positively and negatively charged particles (i.e., electrons and ions) which are guided by the field line. It is well established that a charged particle in a magnetic field gyrates about field lines, the center of gyration at any instance being called the "guiding center" of the particle. As the gyrating particle moves along a field line in a uniform field, it will follow a helical path about its guiding center, hence linear motion, and will remain on the field line. Electrons and ions both follow helical paths around a field line but rotate in opposite directions. The frequencies at which the electrons and ions rotate about the field line are called gyromagnetic frequencies or cyclotron frequencies because they are identical with the expression for the angular frequencies of gyration of particles in a cyclotron. The cyclotron frequency of ions in a given magnetic field is less than that of electrons, in inverse proportion to their masses.

If the particles which form the plasma along the earth's field lines continued to move with a constant pitch angle, often designated "alpha", they would soon impact on the earth's surface. Pitch angle alpha is defined as the angle between the direction of the earth's magnetic field and the velocity (V) of the particle. However, in converging force fields, the pitch angle does change in such a way as to allow the particle to turn around and avoid impact. Consider a particle moving along a field line down toward the earth. It moves into a region of increasing magnetic field strength and therefore sine alpha increases. But sine alpha can only increase to 1.0, at which point, the particle turns around and starts moving up along the field line, and alpha decreases. The point at which the particle turns around is called the mirror point, and there alpha equals ninety degrees. This process is repeated at the other end of the field line where the same magnetic field strength value B, namely Bm, exists. The particle again turns around and this is called the "conjugate point" of the original mirror point. The particle is therefore trapped and bounces between the two magnetic mirrors. The particle can continue oscillating in space in this manner for long periods of time. The actual place where a particle will mirror can be calculated from the following:

$$\sin^2 \text{alpha}_o = B_o/B_m \quad (1)$$

wherein:

alpha_o = equatorial pitch angle of particle

B_o = equatorial field strength on a particular field line

B_m = field strength at the mirror point

Recent discoveries have established that there are substantial regions of naturally trapped particles in space which are commonly called "trapped radiation belts". These belts occur at altitudes greater than about 500 km and accordingly lie in the magnetosphere and mostly above the ionosphere.

The ionosphere, while it may overlap some of the trapped-particle belts, is a region in which hydrostatic forces govern its particle distribution in the gravitational field. Particle motion within the ionosphere is governed by both hydrodynamic and electrodynamic forces. While there are few trapped particles in the ionosphere, nevertheless, plasma is present along field lines in the ionosphere. The charged particles which form this plasma move between collisions with other particles along similar helical paths around the field lines and although a particular particle may diffuse downward into the earth's lower atmosphere or lose energy and diverge from its original field line due to collisions with other particles, these charged particles are normally replaced by other available charged particles or by particles that are ionized by collision with said particle. The electron density (N_e) of the plasma will vary with the actual conditions and locations involved. Also, neutral particles, ions, and electrons are present in proximity to the field lines.

The production of enhanced ionization will also alter the distribution of atomic and molecular constituents of the atmosphere, most notably through increased atomic nitrogen concentration. The upper atmosphere is normally rich in atomic oxygen (the dominant atmospheric constituent above 200 km altitude), but atomic nitrogen is normally relatively rare. This can be expected to manifest itself in increased airglow, among other effects.

As known in plasma physics, the characteristics of a plasma can be altered by adding energy to the charged particles or by ionizing or exciting additional particles to increase the density of the plasma. One way to do this is by heating the plasma which can be accomplished in different ways, e.g., ohmic, magnetic compression, shock waves, magnetic pumping, electron cyclotron resonance, and the like.

Since electron cyclotron resonance heating is involved in the present invention, a brief discussion of same is in order. Increasing the energy of electrons in a plasma by invoking electron cyclotron resonance heating, is based on a principle similar to that utilized to accelerate charged particles in a cyclotron. If a plasma is confined by a static axial magnetic field of strength B, the charged particles will gyrate about the lines of force with a frequency given, in hertz, as $f_g = 1.54 \times 10^3 B/A$, where: B = magnetic field strength in gauss, and A = mass number of the ion.

Suppose a time-varying field of this frequency is superimposed on the static field B confining the plasma, by passage of a radiofrequency current through a coil which is concentric with that producing the axial field, then in each half-cycle of their rotation about the field lines, the charged particles acquire energy from the oscillating electric field associated with the radio frequency. For example, if B is 10,000 gauss, the frequency of the field which is in resonance with protons in a plasma is 15.4 megahertz.

As applied to electrons, electron cyclotron resonance heating requires an oscillating field having a definite frequency determined by the strength of the confining field. The radio-frequency radiation produces time-varying fields (electric and magnetic), and the electric field accelerates the charged particle. The energized electrons share their energy with ions and neutrals by undergoing collisions with these particles, thereby effectively raising the temperature of the electrons, ions, and neutrals. The apportionment of energy among these species is determined by collision frequencies. For a more detailed understanding of the physics involved, see "Controlled Thermonuclear Reactions", Glasstone and Lovberg, D. Van Nostrand Company, Inc., Princeton, N.J., 1960 and "The Radiation Belt and Magnetosphere", Hess, Blaisdell Publishing Company, 1968, both of which are incorporated herein by reference.

Referring now to the drawings, the present invention provides a method and apparatus for altering at least one region of plasma which lies along a field line, particularly when it passes through the ionosphere and/or magnetosphere. FIG. **1** is a simplified illustration of the earth **10** and one of its dipole magnetic force or field lines **11**. As will be understood, line **11** may be any one of the numerous naturally existing field lines and the actual geographical locations **13** and **14** of line **11** will be chosen based on a particular operation to be carried out. The actual locations at which field lines intersect the earth's surface is documented and is readily ascertainable by those skilled in the art.

Line **11** passes through region R which lies at an altitude above the earth's surface. A wide range of altitudes are useful given the power that can be employed by the practice of this invention. The electron cyclotron resonance heating effect can be made to act on electrons anywhere above the surface of the earth. These electrons may be already present in the atmosphere, ionosphere, and/or magnetosphere of the earth, or can be artificially generated by a variety of means such as x-ray beams, charged particle beams, lasers, the plasma sheath surrounding an object such as a missile or meteor, and the like. Further, artificial particles, e.g., electrons, ions, etc., can be injected directly into region R from an earth-launched rocket or orbiting satellite carrying, for example, a payload of radioactive beta-decay material; alpha emitters; an electron accelerator; and/or ionized gases such as hydrogen; see U.S. Pat. No. 4,042,196. The altitude can be greater than about 50 km if desired, e.g., can be from about 50 km to about 800 km, and, accordingly may lie in either the ionosphere or the magnetosphere or both. As explained above, plasma will be present along line **11** within region R and is represented by the helical line **12**. Plasma **12** is comprised of charged particles (i.e., electrons and ions) which rotate about opposing helical paths along line **11**.

Antenna **15** is positioned as close as is practical to the location **14** where line **11** intersects the earth's surface. Antenna **15** may be of any known construction for high directionality, for example, a phased array, beam spread angle (θ) type. See "The MST Radar at Poker Flat, Alaska", Radio Science, Vol. 15, No. 2, Mar.-Apr. 1980, pps. 213–223, which is incorporated herein by reference. Antenna **15** is coupled to transmitter **16** which generates a beam of high frequency electromagnetic radiation at a wide range of discrete frequencies, e.g., from about 20 to about 1800 kilohertz (kHz).

Transmitter **16** is powered by power generator means **17** which is preferably comprised of one or more large, commercial electrical generators. Some embodiments of the present invention require large amounts of power, e.g., up to 10^9 to 10^{11} watts, in continuous wave or pulsed power. Generation of the needed power is within the state of the art. Although the electrical generators necessary for the practice of the invention can be powered in any known manner, for example, by nuclear reactors, hydroelectric facilities, hydrocarbon fuels, and the like, this invention, because of its very large power requirement in certain applications, is particularly adapted for use with certain types of fuel sources which naturally occur at strategic geographical locations around the earth. For example, large reserves of hydrocarbons (oil and natural gas) exist in Alaska and Canada. In northern Alaska, particularly the North Slope region, large reserves are currently readily available. Alaska and northern Canada also are ideally located geographically as to magnetic latitudes. Alaska provides easy access to magnetic field lines that are especially suited to the practice of this invention, since many field lines which extend to desirable altitudes for this invention intersect the earth in Alaska. Thus, in Alaska, there is a unique combination of large, accessible fuel sources at desirable field line intersections. Further, a particularly desirable fuel source for the generation of very large amounts of electricity is present in Alaska in abundance, this source being natural gas. The presence of very large amounts of clean-burning natural gas in Alaskan latitudes, particularly on the North Slope, and the availability of magnetohydrodynamic (MHD), gas turbine, fuel cell, electrogasdynamic (EGD) electric generators which operate very efficiently with natural gas provide an ideal power source for the unprecedented power requirements of certain of the applications of this invention. For a more detailed discussion of the various means for generating electricity from hydrocarbon fuels, see "Electrical Aspects of Combustion", Lawton and Weinberg, Clarendon Press, 1969. For example, it is possible to generate the electricity directly at the high frequency needed to drive the antenna system. To do this, typically the velocity of flow of the combustion gases (v), past magnetic field perturbation of dimension d (in the case of MHD), follow the rule:

$$v = df$$

where f is the frequency at which electricity is generated. Thus, if $v = 1.78 \times 10^6$ cm/sec and d=1 cm then electricity would be generated at a frequency of 1.78 mHz.

Put another way, in Alaska, the right type of fuel (natural gas) is naturally present in large amounts and at just the right magnetic latitudes for the most efficient practice of this invention, a truly unique combination of circumstances. Desirable magnetic latitudes for the practice of this invention interest the earth's surface both northerly and southerly of the equator, particularly desirable latitudes being those, both northerly and southerly, which correspond in magnitude with the magnetic latitudes that encompass Alaska.

Referring now to FIG. **2** a first ambodiment is illustrated where a selected region R_1 of plasma **12** is altered by electron cyclotron resonance heating to accelerate the electrons of plasma **12**, which are following helical paths along field line **11**.

To accomplish this result, electromagnetic radiation is transmitted at the outset, essentially parallel to line **11** via antenna **15** as right hand circularly polarized radiation wave **20**. Wave **20** has a frequency which will excite electron cyclotron resonance with plasma **12** at its initial or original altitude. This frequency will vary depending on the electron cyclotron resonance of region R_1 which, in turn, can be determined from available data based on the altitudes of region R_1, the particular field line **11** being used, the strength of the earth's magnetic field, etc. Frequencies of from about 20 to about 7200 kHz, preferably from about 20 to about 1800 kHz can be employed. Also, for any given application, there will be a threshhold (minimum power level) which is needed to produce the desired result. The minimum power level is a function of the level of plasma production and movement required, taking into consideration any loss processes that may be dominant in a particular plasma or propagation path.

As electron cyclotron resonance is established in plasma **12**, energy is transferred from the electromagnetic radiation **20** into plasma **12** to heat and accelerate the electrons therein and, subsequently, ions and neutral particles. As this process continues, neutral particles which are present within R_1 are ionized and absorbed into plasma **12** and this increases the electron and ion densities of plasma **12**. As the electron energy is raised to values of about 1 kilo electron volt (kev), the generated mirror force (explained below) will direct the excited plasma **12** upward along line **11** to form a plume R_2 at an altitude higher than that of R_1.

Plasma acceleration results from the force on an electron produced by a nonuniform static magnetic field ($\overline{B}$). The force, called the mirror force, is given by

$$F = -\mu \nabla B \qquad (2)$$

where μ is the electron magnetic moment and $\nabla \overline{B}$ is the gradient of the magnetic field, μ being further defined as:

$$W_\perp / B = mV_\perp^2 / 2B$$

where $W_\perp$ is the kinetic energy in the direction perpendicular to that of the magnetic field lines and B is the magnetic field strength at the line of force on which the guiding center of the particle is located. The force as represented by equation (2) is the force which is responsible for a particle obeying equation (1).

Since the magnetic field is divergent in region R_1, it can be shown that the plasma will move upwardly from the heating region as shown in FIG. 1 and further it can be shown that

$$\tfrac{1}{2}M_eV_{e\perp}^2(x) \approx \tfrac{1}{2}M_eV_{e\perp}^2(Y) + \tfrac{1}{2}M_iV_{i\parallel}^2(Y) \qquad (3)$$

where the left hand side is the initial electron transverse kinetic energy; the first term on the right is the transverse electron kinetic energy at some point (Y) in the expanded field region, while the final term is the ion kinetic energy parallel to B at point (Y). This last term is what constitutes the desired ion flow. It is produced by an electrostatic field set up by electrons which are accelerated according to Equation (2) in the divergent field region and pulls ions along with them. Equation (3) ignores electron kinetic energy parallel to B because $V_{e\parallel} \approx V_{i\parallel}$, so the bulk of parallel kinetic energy resides in the ions because of their greater masses. For example, if an electromagnetic energy flux of from about 1 to about 10 watts per square centimeter is applied to region R, whose altitude is 115 km, a plasma having a density (N_e) of 10^{12} per cubic centimeter will be generated and moved upward to region R_2 which has an altitude of about 1000 km. The movement of electrons in the plasma is due to the mirror force while the ions are moved by ambipolar diffusion (which results from the electrostatic field). This effectively "lifts" a layer of plasma **12** from the ionosphere and/or magnetosphere to a higher elevation R_2. The total energy required to create a plasma with a base area of 3 square kilometers and a height of 1000 km is about 3×10^{13} joules.

FIG. **3** is an idealized representation of movement of plasma **12** upon excitation by electron cyclotron resonance within the earth's divergent force field. Electrons (e) are accelerated to velocities required to generate the necessary mirror force to cause their upward movement. At the same time neutral particles (n) which are present along line **11** in region $R_1$are ionized and become part of plasma **12**. As electrons (e) move upward along line **11**, they drag ions (i) and neutrals (n) with them but at an angle θ of about 13 degrees to field line **11**. Also, any particulates that may be present in region R_1, will be swept upwardly with the plasma. As the charged particles of plasma **12** move upward, other particles such as neutrals within or below R_1, move in to replace the upwardly moving particles. These neutrals, under some conditions, can drag with them charged particles.

For example, as a plasma moves upward, other particles at the same altitude as the plasma move horizontally into the region to replace the rising plasma and to form new plasma. The kinetic energy developed by said other particles as they move horizontally is, for example, on the same order of magnitude as the total zonal kinetic energy of stratospheric winds known to exist.

Referring again to FIG. **2**, plasma **12** in region R_1 is moved upward along field line **11**. The plasma **12** will then form a plume (cross-hatched area in FIG. **2**) which will be relatively stable for prolonged periods of time. The exact period of time will vary widely and be determined by gravitational forces and a combination of radiative and diffusive loss terms. In the previous detailed example, the calculations were based on forming a plume by producing 0+energies of 2 ev/particle. About 10 ev per particle would be required to expand plasma **12** to apex point C (FIG. **1**). There at least some of the particles of plasma **12** will be trapped and will oscillate between mirror points along field line **11**. This oscillation will then allow additional heating of the trapped plasma **12** by stochastic heating which is associated with trapped and oscillating particles. See "A New Mechanism for Accelerating Electrons in the Outer Ionosphere" by R. A. Helliwell and T. F. Bell, Journal of Geophysical Research, Vol. 65, No. 6, June, 1960. This is preferably carried out at an altitude of at least 500 km.

The plasma of the typical example might be employed to modify or disrupt microwave transmissions of satellites. If less than total black-out of transmission is desired (e.g., scrambling by phase shifting digital signals), the density of the plasma (N_e) need only be at least about 10^6 per cubic centimeter for a plasma orginating at an altitude of from about 250 to about 400 km and accordingly less energy (i.e., electromagnetic radiation), e.g., 10^8 joules need be provided. Likewise, if the density $N_e$is on the order of 10^8, a properly positioned plume will provide a reflecting surface for VHF waves and can be used to enhance, interfere with, or otherwise modify communication transmissions. It can be seen from the foregoing that by appropriate application of various aspects of this invention at strategic locations and with adequate power sources, a means and method is provided to cause interference with or even total disruption of communications over a very large portion of the earth. This invention could be employed to disrupt not only land based communications, both civilian and military, but also airborne communications and sea communications (both surface and subsurface). This would have significant military implications, particularly as a barrier to or confusing factor for hostile missiles or airplanes. The belt or belts of enhanced ionization produced by the method and apparatus of this invention, particularly if set up over Northern Alaska and Canada, could be employed as an early warning device, as well as a communications disruption medium. Further, the simple ability to produce such a situation in a practical time period can by itself be a deterring force to hostile action. The ideal combination of suitable field lines intersecting the earth's surface at the point where substantial fuel sources are available for generation of very large quantitities of electromagnetic power, such as the North Slope of Alaska, provides the wherewithal to accomplish the foregoing in a practical time period, e.g., strategic requirements could necessitate achieving the desired altered regions in time periods of two minutes or less and this is achievable with this invention, especially when the combination of natural gas and magnetohydrodynamic, gas turbine, fuel cell and/or EGD electric generators are employed at the point where the useful field lines intersect the earth's surface. One feature of this invention which satisfies a basic requirement of a weapon system, i.e., continuous checking of operability, is that small amounts of power can be generated for operability checking purposes. Further, in the exploitation of this invention, since the main electromagnetic beam which generates the enhanced ionized belt of this invention can be modulated itself and/or one or more additional electromagnetic radiation waves can be impinged on the ionized region formed by this invention as will be described in greater detail herein after with respect to FIG. **4**, a substantial amount of randomly modulated signals of very large power magnitude can be generated in a highly nonlinear mode. This can cause confusion of or interference with or even complete disruption of guidance systems employed by even the most sophisticated of airplanes and missiles. The ability to employ and transmit over very wide areas of the earth a plurality of electromagnetic waves of varying frequencies and to change same at will in a random manner, provides a unique ability to interfere with all modes of communications, land, sea, and/or air, at the same time. Because of the unique juxtaposition of usable fuel source at the point where desirable field lines intersect the earth's surface, such wide ranging and complete communication interference can be achieved in a resonably short period of time. Because of the mirroring phenomenon discussed hereinabove, it can also be prolonged for substantial time periods so that it would not be a mere transient effect that could simply be waited out by an opposing force. Thus, this invention provides the ability to put unprecedented amounts of power in the earth's atmosphere at strategic locations and to maintain the power injection level, particularly if random pulsing is employed, in a manner far more precise and better controlled than heretofore accomplished by the prior art, particularly by the detonation of nuclear devices of various yeilds at various altitudes. Where the prior art approaches yielded merely transitory effects, the unique combination of fuel and desirable field lines at the point where the fuel occurs allows the establishment of, compared to prior art approaches, precisely controlled and long-lasting effects which cannot, practically speaking, simply be waited out. Further, by knowing the frequencies of the various electromagnetic beams employed in the practice of this invention, it is possible not only to interfere with third party communications but to take advantage of one or more such beams to carry out a communications network even though the rest of the world's communications are disrupted. Put another way, what is used to disrupt another's communications can be employed by one knowledgeable of this invention as a communications network at the same time. In addition, once one's own communication network is established, the far-reaching extent of the effects of this invention could be employed to pick up communication signals of other for intelligence purposes. Thus, it can be seen that the disrupting effects achievable by this invention can be employed to benefit by the party who is practicing this invention since knowledge of the various electromagnetic waves being employed and how they will vary in frequency and magnitude can be used to an advantage for positive communication and eavesdropping purposes at the same time. However, this invention is not limited to locations where the fuel source naturally exists or where desirable field lines naturally intersect the earth's surface. For example, fuel, particularly hydrocarbon fuel, can be transported by pipeline and the like to the location where the invention is to be practiced.

FIG. **4** illustrates another embodiment wherein a selected region of plasma R_3which lies within the earth's ionosphere is altered to increase the density thereof whereby a relatively stable layer **30** of relatively dense plasma is maintained within region R_3. Electromagnetic radiation is transmitted at the outset essentially parallel to field line **11** via antenna **15** as a right hand circularly polarized wave and at a frequency (e.g., 1.78 megahertz when the magnetic field at the desired altitude is 0.66 gauss) capable of exciting electron cyclotron resonance in plasma **12** at the particular altitude of plasma **12**. This causes heating of the particles (electrons, ions, neutrals, and particulates) and ionization of the uncharged particles adjacent line **11**, all of which are absorbed into plasma **12** to increase the density thereof. The power transmitted, e.g., 2×10^6 watts for up to 2 minutes heating time, is less than that required to generate the mirror force F required to move plasma **12** upward as in the previous embodiment.

While continuing to transmit electromagnetic radiation **20** from antenna **15**, a second electromagnetic radiation beam **31**, which is at a defined frequency different from the radiation from antenna **15**, is transmitted from one or more second sources via antenna **32** into layer **30** and is absorbed into a portion of layer **30** (cross-hatched area in FIG. **4**). The electromagnetic radiation wave from antenna **32** is amplitude modulated to match a known mode of oscillation f_3 in layer **30**. This creates a resonance in layer **30** which excites a new plasma wave **33** which also has a frequency of f_3 and which then propogates through the ionosphere. Wave **33** can be used to improve or disrupt communications or both depending on what is desired in a particular application. Of course, more than one new wave **33** can be generated and the various new waves can be modulated at will and in a highly nonlinear fashion.

FIG. **5** shows apparatus useful in this invention, particularly when those applications of this invention are employed which require extremely large amounts of power. In FIG. **5** there is shown the earth's surface **40** with a well **41** extending downwardly thereinto until it penetrates hydrocarbon producing reservoir **42**. Hydrocarbon reservoir **42** produces natural gas alone or in combination with crude oil. Hydrocarbons are produced from reservoir **42** through well **41** and wellhead **43** to a treating system **44** by way of pipe **45**. In treater **44**, desirable liquids such as crude oil and gas condensates are separated and recovered by way of pipe **46** while undesirable gases and liquids such as water, H_2S, and the like are separated by way of pipe **47**. Desirable gases such as carbon dioxide are separated by way of pipe **48**, and the remaining natural gas stream is removed from treater **44** by way of pipe **49** for storage in conventional tankage means (not shown) for future use and/or use in an electrical generator such as a magnetohydrodynamic, gas turbine, fuel cell or EGD generator **50**. Any desired number and combination of different types of electric generators can be employed in the practice of this invention. The natural gas is burned in generator **50** to produce substantial quantities of electricity which is then stored and/or passed by way of wire **51** to a transmitter **52** which generates the electromagnetic radiation to be used in the method of this invention. The electromagnetic radiation is then passed by way of wire **53** to antenna **54** which is located at or near the end of field line **11**. Antenna **54** sends circularly polarized radiation wave **20** upwards along field line **11** to carry out the various methods of this invention as described hereinabove.

Of course, the fuel source need not be used in its naturally-occurring state but could first be converted to another second energy source form such as hydrogen, hydrazine and the like, and electricity then generated from said second energy source form.

It can be seen from the foregoing that when desirable field line **11** intersects earth's surface **40** at or near a large naturally-occurring hydrocarbon source **42**, exceedingly large amounts of power can be very efficiently produced and transmitted in the direction of field lines. This is particularly so when the fuel source is natural gas and magnetohydrodynamic generators are employed. Further, this can all be accomplished in a relatively small physical area when there is the unique coincidence of fuel source **42** and desirable field line **11**. Of course, only one set of equipment is shown in FIG. **5** for sake of simplicity. For a large hydrocarbon reservoir **42**, a plurality of wells **41** can be employed to feed one or more storage means and/or treaters and as large a number of generators **55** as needed to power one or more transmitters **52** and one or more antennas **54**. Since all of the apparatus **44** through **54** can be employed and used essentially at the sight where naturally-occurring fuel source **42** is located, all the necessary electromagnetic radiation **20** is generated essentially at the same location as fuel source **42**. This provides for a maximum amount of usable electromagnetic radiation **20** since there are no significant storage or transportation losses to be incurred. In other words, the apparatus is brought to the sight of the fuel source where desirable field line **11** intersects the earth's surface **40** on or near the geographical location of fuel source **42**, fuel source **42** being at a desirable magnetic latitude for the practice of this invention, for example, Alaska.

The generation of electricity by motion of a conducting fluid through a magnetic field, i.e., magnetohydrodynamics (MHD), provides a method of electric power generation without moving mechanical parts and when the conducting fluid is a plasma formed by combustion of a fuel such as natural gas, an idealized combination of apparatus is realized since the very clean-burning natural gas forms the conducting plasma in an efficient manner and the thus formed plasma, when passed through a magnetic field, generates electricity in a very efficient manner. Thus, the use of fuel source **42** to generate a plasma by combustion thereof for the generation of electricity essentially at the site of occurrence of the fuel source is unique and ideal when high power levels are required and desirable field lines **11** intersect the earth's surface **40** at or near the site of fuel source **42**. A particular advantage for MHD generators is that they can be made to generate large amounts of power with a small volume, light weight device. For example, a 1000 megawatt MHD generator can be construed using superconducting magnets to weigh roughly 42,000 pounds and can be readily air lifted.

This invention has a phenomenal variety of possible ramifications and potential future developments. As alluded to earlier, missile or aircraft destruction, deflection, or confusion could result, particularly when relativistic particles are employed. Also, large regions of the atmosphere could be lifted to an unexpectedly high altitude so that missiles encounter unexpected and unplanned drag forces with resultant destruction or deflection of same. Weather modification is possible by, for example, altering upper atmosphere wind patterns or altering solar absorption patterns by constructing one or more plumes of atmospheric particles which will act as a lens or focusing device. Also as alluded to earlier, molecular modifications of the atmosphere can take place so that positive environmental effects can be achieved. Besides actually changing the molecular composition of an atmospheric region, a particular molecule or molecules can be chosen for increased presence. For example, ozone, nitrogen, etc. concentrations in the atmosphere could be artificially increased. Similarly, environmental enhancement could be achieved by causing the breakup of various chemical entities such as carbon dioxide, carbon monoxide, nitrous oxides, and the like. Transportation of entities can also be realized when advantage is taken of the drag effects caused by regions of the atmosphere moving up along diverging field lines. Small micron sized particles can be then transported, and, under certain circumstances and with the availability of sufficient energy, larger particles or objects could be similarly affected. Particles with desired characteristics such as tackiness, reflectivity, absorptivity, etc., can be transported for specific purposes or effects. For example, a plume of tacky particles could be established to increase the drag on a missile or satellite passing therethrough. Even plumes of plasma having substantially less charged particle density than described above will produce drag effects on missiles which will affect a lightweight (dummy) missile in a manner substantially different than a heavy (live) missile and this affect can be used to distinguish between the two types of missiles. A moving plume could also serve as a means for supplying a space station or for focusing vast amount of sunlight on selected portions of the earth. Surveys of global scope could also be realized because the earth's natural magnetic field could be significantly altered in a controlled manner by plasma beta effects resulting in, for example, improved magnetotelluric surveys. Electromagnetic pulse defenses are also possible. The earth's magnetic field could be decreased or disrupted at appropriate altitudes to modify or eliminate the magnetic field in high Compton electron generation (e.g., from high altitude nuclear bursts) regions. High intensity, well controlled electrical fields can be provided in selected locations for various purposes. For example, the plasma sheath surrounding a missile or satellite could be used as a trigger for activating such a high intensity field to destroy the missile or satellite. Further, irregularities can be created in the ionosphere which will interfere with the normal operation of various types of radar, e.g., synthetic aperture radar. The present invention can also be used to create artificial belts of trapped particles which in turn can be studied to determine the stability of such parties. Still further, plumes in accordance with the present invention can be formed to simulate and/or perform the same functions as performed by the detonation of a "heave" type nuclear device without actually having to detonate such a device. Thus it can be seen that the ramifications are numerous, far-reaching, and exceedingly varied in usefulness.

I claim:

1. A method for altering at least one region normally existing above the earth's surface with electromagnetic radiation using naturally-occurring and diverging magnetic field lines of the earth comprising transmitting first electromagnetic radiation at a frequency between 20 and 7200 kHz from the earth's surface, said transmitting being conducted essentially at the outset of transmission substantially parallel to and along at least one of said field lines, adjusting the frequency of said first radiation to a value which will excite electron cyclotron resonance at an initial elevation at least 50 km above the earth's surface, whereby in the region in which said electron cyclotron resonance takes place heating, further ionization, and movement of both charged and neutral particles is effected, said cyclotron resonance excitation of said region is continued until the electron concentration of said region reaches a value of at least 10^6 per cubic centimeter and has an ion energy of at least 2 ev.

2. The method of claim **1** including the step of providing artificial particles in said at least one region which are excited by said electron cyclotron resonance.

3. The method of claim 2 wherein said artificial particles are provided by injecting same into said at least one region from an orbiting satellite.

4. The method of claim 1 wherein said threshold excitation of electron cyclotron resonance is about 1 watt per cubic centimeter and is sufficient to cause movement of a plasma region along said diverging magnetic field lines to an altitude higher than the altitude at which said excitation was initiated.

5. The method of claim 4 wherein said rising plasma region pulls with it a substantial portion of neutral particles of the atmosphere which exist in or near said plasma region.

6. The method of claim 1 wherein there is provided at least one separate source of second electromagnetic radiation, said second radiation having at least one frequency different from said first radiation, impinging said at least one second radiation on said region while said region is undergoing electron cyclotron resonance excitation caused by said first radiation.

7. The method of claim 6 wherein said second radiation has a frequency which is absorbed by said region.

8. The method of claim 6 wherein said region is plasma in the ionosphere and said second radiation excites plasma waves within said ionosphere.

9. The method of claim 8 wherein said electron concentration reaches a value of at least 10^{12} per cubic centimeter.

10. The method of claim 8 wherein said excitation of electron cyclotron resonance is initially carried out within the ionosphere and is continued for a time sufficient to allow said region to rise above said ionosphere.

11. The method of claim 1 wherein said excitation of electron cyclotron resonance is carried out above about 500 kilometers and for a time of from 0.1 to 1200 seconds such that multiple heating of said plasma region is achieved by means of stochastic heating in the magnetosphere.

12. The method of claim 1 wherein said first electromagnetic radiation is right hand circularly polarized in the northern hemisphere and left hand circularly polarized in the southern hemisphere.

13. The method of claim 1 wherein said electromagnetic radiation is generated at the site of a naturally-occurring hydrocarbon fuel source, said fuel source being located in at least one of northerly or southerly magnetic latitudes.

14. The method of claim 13 wherein said fuel source is natural gas and electricity for generating said electromagnetic radiation is obtained by burning said natural gas in at least one of magnetohydrodynamic, gas turbine, fuel cell, and EGD electric generators located at the site where said natural gas naturally occurs in the earth.

15. The method of claim 14 wherein said site of natural gas is within the magnetic latitudes that encompass Alaska.

* * * * *

CHAPTER THREE

Not Your Average Contrails

Space capabilities are integrated with and affect every link in the kill chain.
—*James G. Roche, U.S. Air Force Secretary (2001–2005)*

The traces in the skies are thus a phenomenon that (at least at first) can, and do, point in any direction. Most remarkable of all is the fact that the longer this phenomenon is known, the more complex become the attempts at explanation, even with no new facts. At first, chemtrails were regarded as visible evidence of concerted spraying activity for weather modification purposes. Then many conspiracy theories also began to take as a point of departure the possibility that this was part of a weapons system (employed in tandem with the HAARP atmospheric heater installed in Alaska, which will be the object of discussion in a later chapter), or of an attempt to carry out targeted biological contamination of large tracts of land. The various conspiracy theories have one element in common: they cannot be disproved, even with the best counterarguments, since persuasive evidence against the theory may also be evaluated as a badge of authenticity for the theory.
— Chris Haderer and Peter Hiess, *Geoengineering, Chemtrails, and Climate As Weapon*, 2005

The U.S. Air Force's credibility is further undermined by the fact that the Department of Defense (DoD) explicitly authorizes the dissemination of misleading information in order to protect classified programs. In a supplement to the National Industrial Security Program manual, released in draft form in March 1992, the DoD told contractors how to draft cover stories that "must be believable and cannot reveal any information regarding the true nature of the contract."
— Bill Sweetman, *Aurora: The Pentagon's Secret Hypersonic Spyplane*, 1993

Our post-industrial, post-Atomic Age atmosphere is now being plasma-ized. Unstable, energetic, high-velocity particles are being cycled into our atmosphere from the stratosphere and ionosphere, then activated by microwaves for military-industrial agendas, all in the guise of ameliorating "climate change." The Cold War may be over, but the Cold Warrior mindset is still alive and well.

With the advent of instant international messaging, chemtrail stories are coming hot and heavy, but for the most part people remain oblivious to exactly what is drifting over our heads or raining down on us. Those who never saw the clouds of the 1950s might take a moment to peruse Clifford Carnicom's "Then and Now" (June 28, 2013)[1] and other photographic comparisons on the Internet.

Chemicals have been affecting clouds to one degree or another since industrialization, the Wright brothers, and World War I[2]. When the first rockets shot into the Space Age from Cape Canaveral, Florida, Americans were swept away by Cold War propaganda and oblivious to the fact that a single NASA space shuttle launch produced tons of carbon dioxide:

> 23 tons of harmful particulate matter settle around the launch area each liftoff, and nearly 13 tons of hydrochloric acid kill fish and plants within half a mile of the site . . .Still, the expected impact of spaceflight pales in comparison with the carbon footprint of a commercial airport. Los Angeles International Airport has carbon dioxide emissions of nearly 19,000 tons a month . . .and the 33,000 airplanes that fly in and our of the airport each month emit about 800,000 tons of carbon dioxide.[3]

And as for blaming carbons for "global warming," the real carbon burners are military F-15s, Apache gunships, Abrams tanks, and non-nuclear carriers. In 2006, the U.S. Air Force burned 2.6 billion gallons of jet fuel. According to the EPA, each gallon of gasoline produces 19.4 pounds of CO_2 and diesel 22 pounds.[4]

Today, the new cloud-watching entails being able to discern natural clouds from artificial or *anthropogenic* clouds, plus how *whole weather systems are being manipulated via geoengineering.*

By 2005, Carnicom had gathered enough data to determine that aerosol operations over northern New Mexico were being staged in advance of approaching moisture and at times of low humidity/low ion count, indicating to him that high moisture content/high negative ion count must not be favorable to aerosol operations. He posited that keeping a close ionic count of the lower atmosphere may be why a multitude of advanced sensors—from antenna farms 5,000 miles up filled with infrared and visible wavelength sensors, to weather sensors on the lower rungs of cell towers—are vacuuming up data so computers can constantly adjust differential and predictive meteorological and conductivity models.

Extreme variations in positive and negative ion counts are a clear indication that nature is being seriously tampered with:

---

1 www.carnicominstitute.org/articles/thenandnow.htm

2 See Arthur W. Clayden. *Cloud Studies.* New York: E.P. Dutton and Company, 1927; at plus.google.com/photos/107393796095434664991/albums/5236028370090070321?banner=pwa

3 Prachi Patel-Predd. "A Spaceport for Tree Huggers." *Discover*, December 2007.

4 Will Thomas, "Nano Chemicals," willthomasonline.net/Nano_Chemtrails.html

> Ion concentration is tied in directly to the electrical nature of the atmosphere and the earth . . .[T]he balances of nature are being upset with artificial methods that threaten the viability of life on this planet.[5]

Add to the confusion of clouds the media campaign to redefine contrails with terms like "persistent contrails" and scare the public with "global warming" and "climate change," and it is no wonder that people look up and see what they're told to see. And yet how can it be that the trails laid out like crossword puzzles or strange circular forms spreading out into an overarching pale haze are the contrails and fumes of normal jet traffic?

CONTRAILS

From the ground, we are accustomed to observing contrails in jets' wake. The possibility that the government and its media have not seen fit to tell us that we might be looking at something different should not surprise us. A contrail typically forms *one wingspan* behind the aircraft (the U.S. Air Force distance standard), but photographic and video evidence consistently indicates that a very different aerosol trail—chemtrail, persistent contrail—forms nearer the engines.[6]

Then there is the presence of *persisting spray or "condensation"* which does not indicate the usual water or ice vapor but rather the presence of particulate matter, perhaps chemicals in the jet fuel and/or what the military terms *chaff.* For ice crystals to persist in contrails at -40°C/-40°F, the altitude moisture content (humidity) is crucial, which means that contrails can also occur at low altitudes, especially in the cold winter months. Once airborne particles like barium or aluminum are added to create more nuclei for atmospheric moisture to condense around in low humidity areas, droplets will freeze into persistent plumes at much lower altitudes (and higher temperatures), such as the "persistent contrail" over the Afghani desert that the *New York Times* ran a photograph of in 2011.[7]

Since 1998, air traffic controllers (ATCs) at major airports in the U.S. and Canada have been told to divert commercial traffic away from "commanded airspace" and warned that "climate experiments" might degrade their radar displays. Some ATCs figured out that the artificial stuff obscuring their radar had something to do with hospital emergency rooms being filled to overflowing with acute respiratory attacks. Chemtrail complaints and photographs went

---

5 Clifford E. Carnicom, "Ions and Humidity," May 26, 2005. Carnicom recommends utilizing the two variables of humidity and negative ion count when analyzing aerosol operations in geographic areas.

6 Clifford E. Carnicom, "Contrail Distance Formation Model," March 22, 2001. At "Contrail Formation Model" (April 12, 2001), Carnicom offers a mathematical formula of hard scientific data for contrail formation under various meteorological conditions at various flight altitudes.

7 Eric Schmitt and Tom Shanker, "In Long Pursuit of Bin Laden, the '07 Raid, and Frustration." *New York Times*, May 5, 2011.

viral over the Web, but other than Representative Kucinich's doomed Space Preservation Act of 2001, official silence has prevailed for a decade and a half. In fact, under the cloak of "sequestration," the FAA has now closed 149 air traffic control (ATC) towers:

> The airports targeted for tower shutdowns have fewer than 150,000 total flight operations per year. Of those, fewer than 10,000 are commercial flights by passenger airlines.[8]

Of the 140,000 noncommercial flights, how many are military and military contractor chemtrail flights and UAVs/drones[9] overriding ATC scrutiny? Hundreds of small airports routinely operate without ATCs, which means they are perfect for all kinds of under-the-radar operations. Even the U.S. Forest Service now has drones.[10]

CHEMTRAILS

As was noted in the Introduction, the disparaged term "chemtrails" is a contraction for an aviation smog of "chemical trails" and hearkens from the 1990 U.S. Air Force Academy Chemistry 131 manual preparing cadets in molecular geometry, acid rain, spectroscopy, acid base titration, the chemistry of photography, identification of chemical compounds, chemical kinetics, electrochemistry, and organic chemistry in preparation for Air Force aerosol programs. On the cover by the word "Chemtrails" is a rocket spouting fumes.

Readers may recall how all commercial flights were grounded for some time after the destruction of the World Trade Towers in New York City on September 11, 2001. Thomas relates that only nine selected military tanker planes were in the sky, crossing between Ohio and Virginia—"an area normally cloaked in cirrus clouds from the 700 to 800 jets crossing daily"—and leaving what NASA's contrail expert Dr. Patrick Minnis[11] insists were "threadlike contrails fanned out over five hours to form a shield of cirrus clouds covering 24,000 square miles."[12] Given that real contrails don't "fan out over five hours," sensors may have been about measuring chemtrail distribution without all the usual commercial contrail "noise."

---

8 Jason Keyser, "FAA to close 149 air traffic towers under cuts." AP, March 22, 2013.

9 The military claims it is converting drones from reconnaissance to hurricane trackers for HS3 (Hurricane and Severe Storm Sentinel), but it is more likely that more than a few are being reconfigured for chemtrail missions. See "NASA sends drones to track hurricanes' secrets," *The News*, September 15, 2013, www.thenews.com.pk

10 "Forest Service Buys Flying Drones to Find Marijuana Growers in Calif." AP, April 4, 2008.

11 Minnis, J. Kirk Ayers and Steven P. Weaver coined the term *cirrus contrailus* in their report, "Surface-Based Observations of Contrail Occurrence Frequency Over the U.S., April 1993–April 1994," NASA RP-1404, December 1997.

12 "Briefly Empty Skies Offer Climate Clues." *New York Times*, October 30, 2001.

In fact, it is puzzling that during that September 11–14 window, a survey of the average Diurnal Temperature Range (DTR) indicated a temperature *increase* of 10°C (50°F), and once normal air traffic resumed, a temperature *drop* of 0.8°C (33.44°F).[13]

> The increase is larger than during the 11–14 September period for the previous 30 years, giving ammunition to critics who state that weather conditions at this specific period were very extraordinary and no scientific based conclusions could be taken. However, even more surprising is the fact that the 11–14 September increase in DTR was more than twice the national average for regions in the United States where contrail coverage was previously reported to have been most abundant, such as the Midwest, Northeast, and Northwest regions.[14]

Surprising, indeed.

What is not surprising is that aerosol delivery systems have been all the rage since militaries everywhere took to the air. In the mid-1950s and again a decade later, St. Louis kids watched low-flying Army planes crop-dust zinc cadmium sulfide laced with radioactive particles over their neighborhood.

> . . .the Army used motorized blowers atop a low-income housing high-rise, at schools, and from the backs of station wagons to send a potentially dangerous compound into the already-hazy air in predominantly black areas of St. Louis.[15]

Mary Helen Brindell went inside and washed it off her face and arms, then went back outside to play, not knowing that over time the cadmium would degenerate her breast, thyroid, skin, and uterus.

Decades after the Vietnam "conflict," Americans finally learned about Agent Orange. Recently, a veterans' services officer unearthed records of the 267th Chemical Platoon that in 2000 forced the Pentagon to finally admit to Project 112 and thousands of sub-projects that secretly exposed thousands of American military personnel in Hawaii, Panama, Okinawa, and aboard ships in the Pacific Ocean to poisons, drugs, and germs, including sarin and VX nerve gases (1962–1974).[16]

From secret clinical trials of psychochemicals at Edgewood Arsenal on Chesapeake Bay,[17] to "aerial fumigation" of Monsanto's Roundup herbicide over Colombian subsistence crops, livestock, villages, schools, and churches in the

---

13 Brandon Turbeville, "Belgian Environmental Study Corroborates Existence and Effects of Weather Modification." AP, January 3, 2013.

14 "Case Orange: Contrail Science, Its Impact on Climate and Weather Manipulation Programs Conducted By the United States and Its Allies." Belfort Group, 10 May 2010. www.belfort-group.eu/sites/default/files/page/2010/05/COpart1.pdf

15 Jim Salter, "Secret Cold War tests on St. Louis raise concerns." AP, October 4, 2012.

16 Jon Mitchell, "Were US Marines Used as Guinea Pigs on Okinawa?" *CounterPunch*, December 19, 2012.

17 Raffi Khatchadourian, "Operation Delirium." *The New Yorker*, December 17, 2012.

name of the war on drugs,[18] exploiting uninformed populations has never ended. In 2007, the California Department of Food and Agriculture (CDFA)—endorsed and bankrolled by the EPA, USDA, and Department of Homeland Security (DHS)—sprayed miniature time-release microcapsules filled with Checkmate OLR-F and LBAM-F (synthetic moth sex pheromone) over Monterey, San Mateo, San Francisco, and Oakland, purportedly to eliminate less than 9,000 light brown apple moths—the same moths that New Zealand manages with natural predators and washing fruit with water. The capsules alone contain BHT (butelated hydroxyl tolune), which causes sterility in men, and an endocrine disruptor that forces constant production of estrogen in men, women, and children.

The description of the "experiment" doesn't sound all that different from the chemtrail particulates, polymers, and sensors discussed later in this chapter:

> In military technology, microcapsules can be used for chemical shields *or for experimental 'non-lethal' weapons.* Like pollen, the miniature capsules can float in the air and they can stick to surfaces people touch. Imagine hundreds of synthetic microscopic plastic balls floating in the environment and then entering your nose, mouth, eyes and ears . . .They get lodged in your lungs, you cough profusely to expel them, and you go into respiratory distress.[19] (Emphasis added.)

The three-year program of nine aerial sprayings cost $3.5 million per spray. Suterra LLC of Bend, Oregon raked in the profits while the people below fell ill:

> Never mind that known mutagens, carcinogens, lung, skin and eye irritants, and compounds considered hazardous waste by the European Union, all micronized in plastic capable of remaining lodged in the deep lung, were in the first round of spraying for the LBAM in the Fall of 2007 in Monterey and Santa Cruz counties. Then, hundreds of seabirds died afterwards, accompanied temporarily by one of the worst red tides in 40 years—and hundreds of people had health complaints, even though the CDFA set up no reporting mechanism to monitor these adverse health effects. . .[20]

PROJECT CLOVERLEAF AND THE BIG BUSINESS OF GEOENGINEERING

Geoengineering, the so-called antidote to "global warming," is actually about yet more "experimental" chemical warfare being waged at civilian expense, this time under the rubric of Project Cloverleaf:

18 Jenny O'Connor, "Colombia's Agent Orange?" *CounterPunch*, October 31, 2012.

19 Rami Nagel, "Governor Schwarzenegger Backed Immoral Sex Pheromone Spraying Continues. . ." NaturalNews.com, February 8, 2008.

20 Paulina Borsook, "Bush Administration to Blue-State California: Drop Dead!" March 20, 2008.

> Project Cloverleaf is a joint US-Canadian military operation involving distributing chemicals into the atmosphere above Canada and the United States. Both US military refueling tankers and thousands of planes in private corporate aviation are used. Military & civilian aspects of Project Cloverleaf are covert operations. The purpose is to seed into the atmosphere multiple weather/climate modification chemicals for purposes of proactive environmental warfare, originally motivated by a climate change concern; & to introduce highly humanly toxic metallic salts and aerosol fibers that facilitate atmospheric operations of HAARP technology (which is involved in climate manipulation). Piggybacking on this, the covert distribution framework of the toxic metals & chemicals has been used in other covert military/civilian operations like massive biological experiments on whole cities and countrysides of people/ecologies—tests which are unauthorized & without consent or even public knowledge. The purpose is nothing less than the actual physical transformation of the earth's atmosphere in order to provide a platform for the latest chemical & electromagnetic technologies of warfare, communication, weather control, low-yield biological warfare, and control of populations through "non-lethal" chemical/electromagnetic means. Project Cloverleaf is a global phenomenon. This is its short, documented history.[21]

First, the public groundwork had to be laid for Cloverleaf, beginning with the March 26, 1991 Hughes Aircraft Patent #5,003,186, "Stratospheric Welsbach seeding for reduction of global warming," that detailed the 8–12 micron infrared standard for aluminum oxide and other reflective particulates. HAARP and chemtrails were revving into high gear by August 20–23, 1997 when Manhattan Project father Edward Teller—founder of the Lawrence Livermore National Lab and senior research fellow at the Hoover Institution of War, Revolution and Peace at Stanford University since 1975—presented "Global Warming and Ice Ages: I. Prospects for Physics-Based Modulation of Global Change" to the 22nd International Seminar on Planetary Emergencies in Erice, Italy. Two months later, he followed it up with a disingenuous piece in the *Wall Street Journal* recommending Project Cloverleaf without naming it or admitting that it already existed:

> One particularly attractive approach involves diminishing slightly—by about 1 percent—the amount of sunlight reaching the earth's surface in order to counteract any warming effect of greenhouse gases . . .[I]f the politics of global warming require that "something must be done" while we still don't know whether anything really needs to be done—let alone *what* exactly—let us play to our uniquely American strengths in innovation and technology to offset any global warming by the least costly means possible. While scientists continue

21 "Project Cloverleaf: Timeline, 1994 to the Present." media.portland.indymedia.org/media/2004/03/283390.pdf

> research into any global climatic effects of greenhouse gases, we ought to study ways to offset any possible ill effects. Injecting sunlight-scattering particles into the stratosphere appears to be a promising approach. Why not do that?[22]

Throughout the first decade of the 21st century, Cloverleaf was black-budget[23] bankrolled as a covert military operation seamlessly sold to the public as stratospheric aerosol geoengineering (SAG) and solar radiation management (SRM). At the Council on Foreign Relations (CFR) meeting on May 5, 2008, sulfur dioxide, aluminum oxide dust, self-levitating aerosols, and sulfuric acid were given the go-ahead in the name of SAG/SRM[24], despite the fact that Cloverleaf was (and still is) a joint U.S.-Canada, computer-satellite military operation broadcasting frequencies to lab-readied, micron-sized particulates being dumped daily on NATO populations.

In fact, a labyrinth of government and corporate complicity lies behind Cloverleaf, all twisting science in every which way to cover over the distribution of the plasma miasma we are now breathing and ingesting. For one example among many, there is the DOE's Office of Biological and Environmental Research (BER), which does "research in the areas of climate and environmental sciences and biological systems science."[25] Under BER, the Climate and Environmental Sciences Division (CESD) runs the Atmospheric System Research Program (ASR) whose mission, "in partnership with the Atmospheric Radiation Measurement (ARM) Climate Research Facility (ACRF), is to quantify interactions among aerosols, clouds, precipitation, radiation, dynamics, and thermodynamics." Fixed and mobile ACRF sites are situated at "climatically diverse locations" to oversee "ground-based and airborne field campaigns to target specific atmospheric processes under a diversity of locations and atmospheric conditions." ACRF studies the "aerosol life cycle," including its "optical properties, with emphasis on the formation of secondary organic aerosols and absorbing carbon aerosols" and "direct and indirect effects on radiation."

Contracting with agencies like BER is the usual phalanx of private contractors—like the Aerosol Technology Laboratory at Texas A&M University that serves both public *and* private sectors, including "the U.S. Army, Los Alamos National Laboratory, Rocky Flats Environmental Technology Site, Savannah River National Laboratory, Pacific Northwest National Laboratory, Department of Energy Waste Isolation Pilot Plant, Siemens-Dematic, Kennecott Utah Copper, Canberra Industries Inc., and Andersen Instruments."[26] Savannah River National Lab developed the aerosol-to-liquid particle extraction system (ALPES)

22 Edward Teller, "The Planet Needs A Sunscreen." *Wall Street Journal*, October 17, 1997.

23 For those unfamiliar with what a black budget is and does, see www.washingtonpost.com/wp-srv/special/national/black-budget

24 "Unilateral Geoengineering: Non-technical Briefing Notes for a Workshop At the Council on Foreign Relations," Washington, D.C., May 5, 2008.

25 science.energy.gov/ber/research/cesd/atmospheric-system-research-program/

26 aerosoltechnologylab.tamu.edu/

and the aerosol contaminant extractor (ACE) for "wide application in homeland security and law enforcement. . .An array of units, deployed throughout a public or private facility, could be a vital part of an antiterrorism alert system."[27] The two portable briefcase-sized devices collect aerosol chemical agents, radioactive particles, microorganisms (spores, bacteria, molds, fungi), residuals from explosives, and manufacturing byproducts like lead.[28]

CLOUD SEEDING GOES ELECTROMAGNETIC

Up until recently, weather engineering was confined to localized ice nucleation operations needing only generators, small planes, and silver iodide crystals to coax a little moisture out of a dry sky, deflect storms, or prevent precipitation. Cloud-seeding flares along the wings of small aircraft would spread silver iodide that would then mix with all the other "precipitation enhancement particles" spewed by commercial and military jets.[29] In April 2013, Weather Modification, Inc. listed 37 cloud-seeding operations in 17 states, and 66 such operations in 18 countries.

Despite the public acceptance of cloud seeding, silver iodide is highly toxic to plants and air breathers. The Office of Environmental Health and Safety at UC Berkeley rates it as a Class C non-soluble, inorganic, hazardous chemical, as does the EPA Clean Water Act. Humans absorb silver iodide through nose, lungs, skin, and GI tract, after which GI inflammation, renal and pulmonary lesions, black and blue skin may follow. As Deepwater Chemicals, a manufacturer of silver iodide, puts it in their Material Safety Data Sheet:

> Chronic ingestion of iodides may produce "iodism," which may be manifested by skin rash, running nose, headache and irritation of the mucous membranes. Weakness, anemia, loss of weight and general depression may also occur. Chronic inhalation or ingestion may cause argyria characterized by blue-gray discoloration of the eyes, skin and mucous membranes. Chronic skin contact may cause permanent discoloration of the skin.[30]

Another chemical technique touted as a "weather control powder" is Patent #6315213, filed on June 21, 2000 by Dyn-O-Mat CEO Peter Cordani:

---

27 "SNRL Devices for Collecting Airborne Material Receive Patents." December 1, 2005 press release. srnl.doe.gov/newsroom/2005news/120105.htm

28 SRNL is managed and operated by Westinghouse Savannah River Company (now the Washington Savannah River Company), a subsidiary of the URS Corporation, one of those "everything" megacorporations that antitrust legislation used to protect the public from.

29 The "party line" on cloud seeding (weather modification) can be found under "Aerosols and Precipitation" at www.rap.ucar.edu/hap/themes/seed.php, the National Center for Atmospheric Research sponsored by the National Science Foundation.

30 www.deepwaterchemicals.com/pdf/107.04.pdf

> . . .a method for artificially modifying the weather by seeding rain clouds of a storm with suitable cross-linked aqueous polymer. The polymer is dispersed into the cloud and the wind of the storm agitates the mixture, causing the polymer to absorb the rain. This reaction forms a gelatinous substance which precipitates to the surface below, thus diminishing the cloud's ability to rain.[31]

Touted as being "completely safe," Dyn-O-Mat is invisible to radar.

Along more electromagnetic lines, infrared laser can be pulsed to strip electrons from atoms and convert them into hydroxyl radicals that make sulfur and nitrogen dioxides form particles for moisture to collect around. North American Weather Consultants and Weather Modification, Inc. utilize lasers, and China employed it to produce snow for its Beijing 2008 Olympics.

> . . .rather than seeding the air with crystals delivered by airplanes or artillery rockets, the Swiss, German and French researchers used a laser to generate 220-millijoule pulses within 60 femtoseconds, where one femtosecond is one millionth of one billionth of a second. That's as much power as 1,000 power plants can generate. . .[32]

Criticized for needing extremely high humidity in low temperatures in order to work, the technique nonetheless worked when fired 60 meters above Berlin in autumn:

> Nothing could be seen with the naked eye, but weather LIDAR/LADAR [Light Detection and Ranging / Laser Imaging Detection and Ranging], which uses lasers to measure light scattering in the atmosphere, confirmed that the density and size of water droplets spiked when the laser was fired.[33]

In Abu Dhabi, the capital of United Arab Emirates (UAE), five WEATHERTEC ionizer units with 20 emitters each produced rain, hail, gales, and lightning in the eastern Al Ain region 52 times during the notoriously parched months of July and August 2010. Far cheaper than desalination[34], the giant ionizers costing £7 million each produce negatively charged particles that attach to ever-present desert dust. For the Abu Dhabi project, the emitters were switched on 74 times over a period of 122 days, basically when the atmospheric humidity had reached the required level of 30 percent or more. Convection then carried the ionized

---

31 Julian Siddle, "US makes 'weather control powder'." *BBC News*, 2 August 2001.

32 Jenny Hsu, "Lasers Could Create Clouds, and Perhaps Rain, on Demand." *Popular Science*, May 3, 2010. For photos of lasers in operation, see Becky Ferreira, "Of Course the Future of Weather Control Involves Lasers." *Motherboard*, no date.

33 Colin Baras, "Laser creates clouds over Germany." *New Scientist*, 2 May 2010. LIDAR detects aerosols the same way sonar uses sound pulses or radar uses radio waves.

34 The removal of salt from sea water to make fresh water for human and animal consumption or irrigation.

particles up to where they then attracted water molecules and made clouds heavy with rain.

The $11 million secret project between the Swiss corporation Meteo Systems International and UAE President HH Sheikh Khalifa bin Zayed Al Nahyan has been monitored by the Max Planck Institute of Meteorology, a leader in the study of atmospheric physics. The Abu Dhabi success may represent a benign face of weather modification, but as former Max Planck director Hartmut Grassl said, there are many applications, and drawing rain to dry areas is only one.[35] Besides, precipitation in one place usually means less precipitation in another place, making weather engineering of any kind a *political* issue.

Black budgets and covert military operations like Cloverleaf tend to go hand in hand with an intelligence presence, especially with HAARP-chemtrails weather ionization in place. SciBlue, Inc. has profited from the shape-shifting line between federal and private. SciBlue—listed as a "U.S. Air Force private company" in La Junta, Colorado—was established in 2009 and incorporated in Texas. If it was in Mission, Texas, then SciBlue may have handled "Guided Missile and Space Vehicle Parts and Auxiliary Equipment." In Colorado Springs, SciBlue is a "veteran-owned small business," the veteran being Col. David Kutchinski, CEO of SciBlue and Legacy Technology Holdings and president and "Bill Payer" of World Peace Technologies. Previously, Kutchinski was a lab tech at Argonne National Laboratory (1992–98), then an FBI Special Agent (1980–85), and finally U.S. Army Deputy Commanding Officer of the 10th Special Forces Group (Airborne), 2001–2003. With expertise in "weather modeling" and "integrating new technologies RAPIDLY," Kutchinski may be connected to Sci-Blue Security Systems in Australia.

Internet weather observer *Dutchsinse* adds Aquiess to Sci-Blue:

> Anyone saying it could not be done [weather modification using smaller stations over the United States]—well, it would appear there are already some fairly large companies DOING IT. In April 2012—two companies using RESONANCE to perform weather modification over Texas—confirmed by Aquiess and Sci-Blue.[36]

And from the Aquiess website:

> Over the past ten years the company Aquiess has repeatedly demonstrated this technology to government and humanitarian observer groups. The proprietary weather modification system operates by *utilizing 'resonance' signals to divert*

35 Karen Leigh, "Abu Dhabi-backed scientists create fake rainstorms in $11 m project." ArabianBusiness.com, 3 January 2011.

36 "Using Frequency to Perform Weather Modification Over Texas — Spring 2012 operations — Aquiess and Sci-Blue," *Dutchsinse*, March 9, 2013.

> *oceanic atmospheric rivers*[37] into areas experiencing severe drought. The Aquiess system does not rely on chemical or biologically hazardous materials, which could potentially harm the environment.[38]

The term *atmospheric river* was coined by researchers Reginald Newell and Yong Zhu of MIT in the early 1990s. It is a narrow plume of moisture several thousand kilometers long and a few hundred kilometers wide driven by jet stream winds. Typically, an atmospheric river runs along the boundaries of large areas of divergent surface airflow, including frontal zones associated with extratropical cyclones forming over oceans. One atmospheric river alone carries more water than the Amazon River, and typically there are three to five rivers moving over a hemisphere at any given time. Being able to divert atmospheric rivers spells power over the weather.

Insider rainmaker corporations like SciBlue and Aquiess are about far more than delivering rain.[39] To be blunt, electromagnetic weather engineering that can create and steer extreme weather promises untold profit and power. NASA climate scientist Jim Hansen is even more blunt: "Human-induced climate change is a great moral issue on par with slavery."

So what about the blizzards and freezing temperatures that killed 30,000 UK cows and sheep as more than 5,000 pensioners struggle to stay warm in April 2013?[40] Was it due to "global warming," or is extreme weather being engineered?

THE DELIVERY SYSTEM

At any moment of the day wherever we live, whether or not we see it, the chemtrail delivery system is in full operation. Like the trihalomethanes from disinfectants running down our drains and merging with rotting organic matter[41], not seeing it doesn't mean it's not there, nor does it mean that thousands of people won't sicken and die from what they can't see.

Look up and you might see what looks like eight or more trails from the wing assemblies of one jet and perhaps another trail from the center of the fuselage and tail. Suddenly, the nozzles turn off; farther across the sky, they turn back on.

---

37 See Col. Kutchinski's "Using the Rivers of the Troposphere" at blog.chron.com/climateabyss/2011/07/diary-entry-the-last-word/

38 Also see tatoott1009.wordpress.com/2013/03/10/smoking-gun-on-an-active-program-using-frequency-to-control-the-weather-in-the-continental-united-states/

39 In February 2013, the U.S. Department of Agriculture reported that the drought ravaging 57 percent of the United States has culled Texas cattle to what they were in 1952. Elizabeth Campbell, "Odds Against U.S. Drought Ending Soon, Texas Climatologist Says." *Bloomberg*, February 6, 2013.

40 Richard Gray, "Britain shivers in coldest April day for 20 years." *The Telegraph*, 4 April 2013.

41 In 2011, the watchdog Environmental Working Group examined water quality tests for over two hundred municipal water systems affecting 100 million people in 43 states. See Andrea Germanos, "Study: Over 100 Million Americans Drinking 'Toxic Trash' Water." *Common Dreams*, February 27, 2013.

The multiple trails slowly spread out in an artificial cirrus cloud cover in an already pale sky.

Even at night, trails are being laid. As to why, "Deep Shield," an early chemtrail whistleblower (or disinformation agent), offered the following: "By strategically spraying in certain areas at night, we get the advantage of the rising air, which not only pushes the material [particulates] higher, but also causes the material to disperse into a thin layer."[42] EQUIS II (mentioned in Chapter 1) investigates nighttime plasma structures, electrodynamics, and mesospheric scattering processes, all of which require a distribution of the luminous tracer trimethyl aluminum (TMA) so as to follow the atmospheric motions that form the milky clouds of the nighttime sky.

Cloud physicist William R. Cotton also talked about night flights for mid-level stratus seeding: by day, systematic seeding of non-freezing stratus clouds with pollution aerosols (small hygroscopic particles) will increase albedo (reflected sunlight), but by night, it is more advantageous to seed with large hydroscopic particles.[43] (Given that *hygroscopic* refers to a substance's ability to attract and hold water molecules, and *hydroscopic* to an optical device for viewing objects far below the surface of water, it appears that Cotton either used *hydroscopic* and meant *hygroscopic*, or night runs are about *optics*.)

Below on the ground, big Trimac Western[44] tankers crawl along the highways, hauling witches' brews of chemicals to military and civilian airfields where specialized Boeing 747 Supertankers await them. No Department of Transportation (DOT) placards announce the contents codes of these tankers.

Besides the pivotal Wright-Patterson Air Force Base in Ohio and Kirtland Air Force Base in New Mexico, other military bases engaged in chemtrails delivery include:

George AFB (previously closed), Mojave Desert, California
Nellis AFB and the Mancamp Complex near Tonopah, Nevada
Williams AFB (previously closed), Phoenix, Arizona
Pinal Airpark, Marana, Arizona
NORAD, Peterson AFB, Colorado Springs, Colorado
Air National Guard, Lincoln, Nebraska
Tinker AFB, Oklahoma City, Oklahoma
Fort Sill, Oklahoma
Dobbins Air Base, Marietta, Georgia
McGuire AFB, Trenton, New Jersey
Brunswick Naval Air Station, Maine

---

42 www.holmestead.ca/chemtrails/shieldproject.html

43 March 2008 lecture "Perturbed Clouds in the Climate System" at the Frankfurt Institute for Advanced Studies (FAS) in Frankfurt, Germany.

44 Trimac has tanks, ships, tank farms, rail car tanks, highway tankers and terminals serving BASF, DuPont, Bayer, and other Big Chem. Trimac is owned by UK's Rentokil Initial.

Then there are private cargo carriers like Evergreen International Airlines, a known CIA carrier based outside McMinnville, Oregon near Portland, which appears to be finally shutting down its specially outfitted fleet of 747s. ("The privately held company hasn't disclosed revenues since 2004."[45]) From its website:

> The Evergreen Supertanker is not just limited to fighting fire. It will be a true utilitarian aircraft with the capability to configure to different applications on short notice. This multimission aircraft can support *sensitive security and environmental missions.* The aircraft's exceptional drop capabilities, loiter time and size make it an ideal tool to perform challenging homeland security missions, able to neutralize chemical attacks on military installments or major population centers, and help control large, environmentally disastrous oil spills. [Emphasis added.]

Evergreen's FAA *supplemental* certificate (held since 1975) has been perfect for clandestine operations, given that unlike *flag* certificates, supplementals "ease restrictions":

> . . .a non-scheduled operator can run a charter service, enabling customers to go almost anywhere in the world, to keep records differently and to use alternate manuals for dispatch and communications . . .At one time Evergreen had authority to fly almost anywhere, and it may still.[46]

The status of Evergreen International Airlines' parent company Evergreen International Aviation Inc. is unclear. Is Evergreen's competitor Atlas Air Worldwide Holdings now the CIA's chemtrails cargo carrier of choice?

Jumbo jets can be flown remotely as UAVs or drones, with capacitors making sure that beaming is not continuous. The Abstract of U.S. Patent #5,068,669 "Power Beaming System" lays out the advantages of remote microwave powering:

> Airborne craft powered by microwave beams from the earth's surface will never run out of fuel. Their route may be circumscribed by the pattern of the energizing beams, but they can stay aloft indefinitely. That fact makes microwave-powered aircraft attractive for surveillance, for relaying communication signals, and for atmospheric data gathering. Moreover, unburdened by a crushing weight of fuel, *such aircraft would be able to lift vast payloads into the upper atmosphere. . .*[47] [Emphasis added.]

---

45 Richard Read, "Evergreen International Airlines flies its last flight and parks aircraft, union says." *The Oregonian*, December 3, 2013.

46 Ibid.

47 See T.W.R. East, "A self-steering array for the SHARP microwave-powered aircraft," *IEEE Transactions on Antennas and Propagations*, vol. 30, no. 12, p. 1565, December 1992.

And it does not appear to be just commercial cargo carriers that are contracting with military or intelligence agencies to carry out aerial spraying; passenger jets as well are involved—all of which points to FAA and DOT complicity.

Of course, the *cost* of daily aerial spraying on a global scale is enough to make the average person disbelieve that it could be happening. Jet costs alone sound prohibitive. A new 747 costs $300 million, not counting fuel and equipment, a large hangar, maintenance, mechanics, flight crew, landing fees, insurance and liability, away expenses, pensions, depreciation—and that's not counting health and environment costs. Pilot Barry Davis ballparks the cost of megasprayers like 747s, 767s, L1011s, DC10s, and MD80s (McDonnell Douglas DC9s) flying at 39,000–45,000 (above commercial flights) at a gallon of fuel per second for climb, cruise, and letdown—a 55-gallon drum per minute. If a 747 lays aerosols for five hours at $4 a gallon, 3,600 gallons per hour, that alone would be $72,000. Fifty aircraft around the globe night and day, 430,000 gallons. In one month, 129 million gallons of fuel at $4/gallon = $578,400,000, and "the whole world wrapped in a cocoon of Chemtrail chemicals, jet exhaust pollution, CO_2, etc., all leading to possible global warming . . .We are spending billions down here on Earth to make our cars and trucks much cleaner, while allowing invisible crap to rain down on us, our Earth and oceans."[48]

Barry's $578,400,000 per month is not that far from the mainstream media's $416.66 million per month for carbon reduction:

> Put into context, the cost of reducing carbon dioxide emissions is currently estimated to be between 0.2 and 2.5 per cent of GDP in the year 2030, which is equivalent to roughly USD $200 to $2000 billion.[49]

Then there are the psychological costs of pilots and air traffic control staff ensnared in these covert operations. For one, pilots flying these aircraft are expected to maintain silence. Here is how an anonymous airline manager describes it:

> . . .They told us that the government was going to pay our airline, along with others, to release special chemicals from commercial aircraft. When asked what the chemicals were and why we were going to spray them, they told us that information was given on a need-to-know basis and we weren't cleared for it. They then went on to state that the chemicals were harmless, but the program was of such importance that it needed to be done at all costs. When we asked them why didn't they just rig military aircraft to spray these chemicals, they

---

48 "Chemtrails are likely to track armies, move viruses around," beforeitsnews.com/chemtrails/2011/10/pilot-speaks-out-about-chemtrails-and-haarp-1256807/html, October 19, 2011.

49 The figure of $5 billion a year was borrowed from the journal *Environmental Research Letters* (3/16/2007). "Shading Earth: Delivering Solar Geoengineering Materials to Combat Global Warming May Be Feasible and Affordable." *Science Daily*, August 29, 2012.

> stated that there weren't enough military aircraft available to release chemicals on such a large basis as needs to be done. That's why Project Cloverleaf was initiated, to allow commercial airlines to assist in releasing these chemicals into the atmosphere. Then someone asked why all the secrecy was needed. The government reps then stated that if the general public knew that the aircraft they were flying on were releasing chemicals into the air, environmentalist groups would raise hell and demand the spraying stop. Someone asked one of the G-men then if the chemicals are harmless, why not tell the public what the chemicals are and why we are spraying them? He seemed perturbed at this question and told us in a tone of authority that the public doesn't need to know what's going on, but that this program is in their best interests. He also stated that we should not tell anyone, nor ask any more questions about it. With that, the briefing was over.[50]

For another psychological (and social) cost, it has been reported that Australian pilots fear "conspiracy theorists" shooting them down. When activists gathered at Sydney Airport to question the pilots involved in spraying, the Australian Federal Police were even on hand:

> "Threats to shoot down aircraft or harming pilots are becoming more prevalent, overt and alarming," says Mike Glynn, a current airline captain and former vice president of the Australian and International Pilots Association. "While most can be discounted as empty threats, there remains the possibility that some unhinged individual or group may act. It is beginning to cause consternation in pilot ranks.[51]

Pilot Glynn then trotted out the usual "water vapour" cover:

> "The upper atmosphere is generally very dry but when conditions permit, such as before the approaching cold front which forces moisture high into the atmosphere, the relative humidity becomes so high that the water vapor in the jet exhaust condenses into ice crystals and a contrail is formed," he said.[52]

But the public's right to the truth is the greatest cost. Few understand that since the National Security Act of 1947, the United States has languished under a news blackout[53] marginalizing everything the military considers "national

---

50 Carnicom, "An airline manager's statement," May 22, 2000. www.carnicominstitute.org/articles/mgr1.htm

51 Peter Holmes, "Jet pilots fear 'chemtrail' attacks." News Limited Network, November 2, 2012.

52 Ibid.

53 Operation Mockingbird: a secret Central Intelligence Agency campaign to influence media beginning in the 1950s. First called Mockingbird in Deborah Davis' 1979 book, *Katharine the Great: Katharine Graham and her Washington Post Empire*. See the 2007 memoir *American Spy: My Secret History in the CIA, Watergate, and Beyond* by convicted Watergate "plumber" E. Howard Hunt; and *The Mighty*

security." Geoengineering articles are being couched in future subjunctive language even as a deafening silence is maintained around HAARP and other ionospheric technologies. As atmospheric scientist Ken Caldeira of the Carnegie Institution for Science puts it: "I hope that we never get to the point where people feel the need to spray aerosols in the sky to offset rampant global warming"[54]; and from that other bastion of objectivity, *Scientific American*, "The white haze that hangs over many major cities could become a familiar sight everywhere if the world decides to try geoengineering to create a cooler planet[55]; and an India Reuters article (making no mention of its own ionospheric heating technology at the National MST Radar Facility at Andra Pradesh and Marion Island meteorological station in the South African Prince Edward Islands in the southern Indian Ocean):

> Planes or airships *could* carry sun-dimming materials high into the atmosphere for an affordable price tag of below $5 billion a year as a way to slow climate change . . .New aircraft, specially adapted to high altitudes, *would* probably be the cheapest delivery system with a price tag of $1 to $2 billion a year, [senior officials meeting in Bangkok this week for a new round of U.N. talks] said. A new hybrid airship *could* be affordable but might be unstable at high altitudes . . .Some experts favour geo-engineering as a quick fix when governments are far from a deal to slow climate change that is expected to cause more heatwaves, floods and rising sea levels . . .Dimming sunlight *would* not, for instance, slow the build-up of carbon dioxide in the atmosphere, which is making the oceans more acidic and undermining the ability of creatures such as mussels or lobsters to build their protective shells.[56] [Emphases added.]

Force pilots to sign confidentiality agreements and employ subjunctives in mainstream media accounts, and you may be able to maintain a secret like Cloverleaf for decades.

Journalists and bloggers who do manage to get a bit of the truth out about what is happening in the skies overhead are quickly discredited or penalized. In 2005 (after the final defeat of the Space Preservation Act), Marcus K. Dalton, managing editor and investigative journalist at the *Las Vegas Tribune*, wrote the two-part "Chemtrails are over Las Vegas":

Wurlitzer: How the CIA Played America by Hugh Wilford (2008).

54 Carnegie Institute, "Geoengineering for Global Warming: Increasing Aerosols in Atmosphere Would Make Sky Whiter." *ScienceDaily*, May 31, 2012.

55 Lauren Morello and ClimateWire, "Geoengineering Could Turn Skies White." *Scientific American*, June 1, 2012.

56 Alister Doyle and David Fogarty, "'Sunshade' to fight climate change cost at $5 bln a year." Reuters India, August 31, 2012.

> . . .Especially disturbing for residents of heavily chemtrailed communities like Las Vegas is a "chemtrail sickness" associated with heavy spray days leaving many stricken people complaining of the "flu" and acute allergic reactions months after the flu season has ended. Upper and lower respiratory and gastrointestinal ailments remain unusually high in many spray areas, along with debilitating fatigue—and something even more worrying.

That was in August. By October, Dalton was looking for a job. Had his publisher been pressured?

A few international incidents have, however, managed to trickle through the "national security" media veil, albeit from "fringe" sources that few Americans access. The first two incidents occurred in India and Nigeria, the last in China.

On June 29, 2009, two Ukrainian AN-124 military cargo aircraft leased and operated by the U.S. Air Force/CIA were forced down in India and Nigeria. China's People's Liberation Army Air Force immediately told Indian and Nigerian intelligence that the U.S. was spraying *biological agents,* possibly H1N1 swine flu virus. India Air Force had been alerted by the AN-124 pilot changing his call signal from civilian to military while preparing to enter Pakistani air space and had forced him to land in Mumbai.

Apparently, both aircraft had waste disposal systems able to hold in excess of 45,000 kg (100,000 pounds). A sophisticated network of "nano-pipes" led to the trailing edges of the wings where horizontal stabilizers dispersed the contents of the waste tanks in an "aerial-type mist." When the U.S. demanded the release of both aircraft, India complied immediately but Nigeria didn't—at least until one of its major oil pipelines was blown up.

There was no mainstream media coverage in the United States, but the small *Fourwinds* article "US Reported in 'Panic' After Chemtrail Planes Forced Down In India, Nigeria" spread like wildfire over the Internet, accompanied by smears of the mysterious "Sorcha Faal Sisters" who had done the reporting. On June 20, 2009, the *Indian Express* reported the same incident, adding more detail *while removing everything regarding the waste disposal system.*[57]

The 2009 downing in China of a Zimbabwean MD-II plane owned by the CIA and Avient Aviation merited similar media confusion. Like Evergreen Airlines, Avient has intelligence ties. In 2003, Avient—supposedly owned by a British military officer named Andrew Smith—was accused by the House of Lords of contracting with Zimbabwe and the Congo to supply military services. In 2006, Avient was accused of bombing civilians in the Congo. In 2008, an Ilyushin Il-76 belonging to Avient delivered 3 million rounds of assault rifle ammunition, 3,000 mortar rounds, and 1,500 rocket-propelled grenades ordered from the Chinese government to Harare[58], thus undermining sanctions against Zimbabwe.

---

57 "Intruder US-hired plane to be let off after Govt clearance." Expressindia.com, June 20, 2009.

58 *The Weekender* (South African newspaper), May 17, 2008; quoted at freebornjohn.blogspot.com/2008/05/Chinese-arms-reach-zimbabwe.html

Avient's intelligence ties makes the downing at Shanghai Pudong Airport on November 28, 2009 less than surprising, as are the polarized Chinese and American accounts of the incident. The "conspiracy theory" version copycatted everywhere on the Internet (again attributed to the marginalized Sorcha Faal Sisters), "CIA Operated Aerial Spraying Plane Shot Down in China" (worldtruth.tv), insisted that Israeli Mossad had done the downing to prevent an American attack upon one of their Central Asian bases in Kyrgyzstan—the "attack" referring to mutated swine flu virus destined for aerial spraying from the Zimbabwean MD-II. The Chinese version made no mention of such illicit cargo and insisted that during takeoff the cargo plane had veered off the runway, caught fire, and crashed—with heavy (chemical) smoke 50 meters high rising into the sky.[59] The day after the crash, xinhuanet.com ran a short piece announcing that the flight data recorder ("black box") had been found.

Bbc.co.uk[60] agreed with both versions that "foreign staff" had been aboard: three killed, four from the U.S., Indonesia, Belgium, and Zimbabwe injured. The "conspiracy theory" version mentioned that the three who died were CIA, and that the Indonesian "confessed" to Chinese secret police that he was a technician working for the U.S. Navy at their Naval Medical Research Unit No. 2 (NAMRU-2) in Indonesia—the very bio-weapons lab that Indonesian defense minister Juwono Sudarsono demanded be closed. The lab is home to the Viral Diseases Program (VDP), "epidemiologic and laboratory research on viral hemorrhagic fevers, influenza, encephalitis, and rickettsioses."[61]

Eleven days after the 2009 downing or crash and a cleanup that might have included swine flu virus, NBC's *Saturday Night Live* ran a tasteless segment about spraying people with viruses from helicopters. The script included such hilarious lines as "Your ears will bleed," "People will get sick as hell—helicopters will spray your asses with virus," and "You will literally be underground."[62]

Now that we have made a few distinctions regarding the difference between contrails and chemtrails and have examined the vast investment that the covert Cloverleaf aerosol delivery system entails, we're ready to take a look at exactly *what* is being sprayed.

---

59 "Three dead, four injured in Cargo plane crash in Shanghai Pudong Airport." xinhuanet.com, November 28, 2009.

60 "Three Americans dead as Zimbabwe plane crashes in China." *BBC News*, 28 November 2009.

61 Paul Watson, "Scientists warn against closing Navy lab." *Los Angeles Times*, July 7, 2008.

62 *Saturday Night Live*, December 6, 2009. I was unable to find a clip.

KUDOS TO THE EUROPEANS

The internal threat for a nation is Lethe, forgetting who we are and why we are here, to lose our bearings, to become anesthetized, to become disillusioned, to rot on the tree that has given us life.

— Georgios Bitros, professor, Economic University, Athens

In *Chemtrails Confirmed,* Will Thomas named 15 nations from which reports of chemtrails were coming: Australia, Belgium, Britain, Canada, Croatia, France, Germany, Holland, Ireland, Italy, New Zealand, Scotland, South Africa, Sweden, and the United States. The most active in Europe to date have been Greece, Cyprus, Germany, Italy, and Spain.

For more than a decade, I have been emailing back and forth with Wayne Hall (www.enouranois.gr/), a Greek citizen born in Australia, graduate of the University of Sydney, teacher and freelance translator in Athens. In the 1980s, Wayne was a member of European Nuclear Disarmament, the non-aligned British-based anti-nuclear-weapons movement, and is a member of the editorial board of the Greek ecological magazine *Nea Ecologia* and founding member of ATTAC-Hellas[63]. Wayne is a pitbull anti-chemtrails activist and has been exceedingly frustrated with Americans who still have not organized to combat chemtrails.[64]

If the primary internal threat to a nation is *lethe* or forgetting, as the opening quote by Georgios Bitros indicates, then the devil stripping nations of their memory will be found in the details, particularly when it comes to piecing together the covert intentions of transnational corporations and governments—or as Andrew Johnson puts it at www.checktheevidence.com, "Any conclusion can be reached about anything—but the value of that conclusion will be proportional to the amount of evidence ignored."

The following timetable/sketch speaks to the consistent activity regarding what is happening in our skies that has been going on in Europe since 1998:

February 9, 1998: Dr. Nick Begich, co-author of *Angels Don't Play This HAARP* (1995), testified before the European Parliament's Committee on Foreign Affairs, Security and Defense Policy about HAARP. (See Appendix F.)

---

63 From www.attac.org/en/overview: "ATTAC is an international organization involved in the alter-globalization movement. We oppose neo-liberal globalization and develop social, ecological, and democratic alternatives so as to guarantee fundamental rights for all. Specifically, we fight for the regulation of financial markets, the closure of tax havens, the introduction of global taxes to finance global public goods, the cancellation of the debt of developing countries, fair trade, and the implementation of limits to free trade and capital flows."

64 See his dialogue with Californian Rosalind Peterson, Agricultural Technologist for the Mendocino County Department of Agriculture (1989–1993) and certified USDA Farm Service Agency Crop Loss Adjustor, also an anti-chemtrails activist: "Dialogue on Persistent Jet Contrails & Experimental Weather Modification," 25 November 2006, www.holmestead.ca/chemtrails/wayne+ros.pdf

> *Motion for Resolution:* "Considers HAARP by virtue of its far-reaching impact on the environment to be a global concern and calls for its legal, ecological and ethical implications to be examined by an international independent body; [the Committee] regrets the repeated refusal of the United States Administration to give evidence to the public hearing into the environmental and public risks of the HAARP program."[65]

November 26, 1998: The European Parliament proposed "a European action plan to integrate military-related resources into environmental strategies [B4-0551/95]."

On May 27, 2000—the day after Croatia joined NATO's "Partnership for Peace"—Croatia's main port town Rijcka was targeted with heavy aerial spraying. Of the more than 300,000 flights that Croatian air control registered, only 25 percent used local airport services, which was authorized by Croatia's Department of Sea, Traffic, and Development. The trail patterns did not match known civilian air corridors and routes. Croatia too experienced an immediate increase in illness.

On Christmas Eve 2002, a huge chemcloud was laid over Aigina, Greece. Contrary to the deafening silence of American mainstream media, the headline in the February 16, 2003 Sunday edition of *Ethnos* trumpeted, "Scientists Uneasy: Dangerous experiments in Greek skies" (by Giannis Kritikos), with the bold subhead, "American aircraft are spraying the atmosphere with chemicals with a view to creating an artificial cloud as an 'antidote' to the Greenhouse Effect." The article included several photographs: "a cloud of chemtrails" over Aigina; a photograph of two Greek scientists discussing "the dangers to public health"; and finally a photograph of a parliamentarian asking, "Who gave permission for this spraying?" A few mornings later, one of the two scientists was addressing the nation on national television.

Given that Greek law states that municipal governments have rights over their land *and* sky, the Aigina Island town council voted to demand an explanation from the Greek government.

In January 2004, Swiss freelance journalist Gabriel Stetter's article "White Skies: The Global Warming Problem and Chemtrails" appeared in the German science magazine *Raum+Zeit* (Space+Time). While furor around the article raged, chemtrail observers joined forces and sought to convince Greenpeace to speak out, but the Greenpeace position to this day has been that they are interested only in issues for which there is "hard data," meaning issues that governments and experts publicly admit to.

In summer 2005, a book by Christian Haderer and Peter Hiess, *Chemtrails: Verschwörung am Himmel (Chemtrails: Conspiracy in the Sky)* came out in German

---

65 Michel Chossudovsky, "It's not only greenhouse gas emissions: Washington's new world order weapons have the ability to trigger climate change." fromthewilderness.com, September 27, 2010.

in Austria. Renamed *Geoengineering, Chemtrails and Climate As A Weapon* when translated into English, it was unfortunately doomed to remain in manuscript, though still available in German on Amazon Germany.[66]

May 10, 2007: Eric Meijer, Dutch Parliamentarian, asked six questions (E-2455/07) of the European Commission, three of which are quoted here:

> "3. Unlike contrails, chemtrails are not an inevitable product of modern aviation. Does the Commission know, therefore, what is the purpose of artificially emitting these Earth-derived substances into the Earth's atmosphere?
>
> "4. To what extent are aerial obscuration and chemtrails now also being employed in the air over Europe, bearing in mind that many people here too are now convinced that the phenomenon is becoming increasingly common and are becoming concerned about the fact that little is so far known about it and there is no public information on the subject? Who initiates this spraying and how is it funded?
>
> "5. Apart from the intended benefits of emitting substances into the air is the Commission aware of any possible disadvantages it may have for the environment, public health, aviation and TV reception?"[67]

June 2007: European Commissioner for the Environment Stavros Dimas responded that there may be risks to environment and health but the Commission saw no reason to act.

December 17, 2007: German Austria admits chemtrail spraying. German Federal Army planes have been spraying at least since March 2006.[68]

Also in 2007, General Fabio Mini, a NATO commander, began writing and speaking out about the new generation of weapons that kill but appear natural, including weather weapons. He should know; he used them during the Kosovo War (February 1998–June 1999), including for artificial cloud cover. He admits that sodium, barium, and aluminum are very useful in the war theater for deflecting electromagnetic waves, and that "chaff" is essential to electronic warfare.[69]

---

66 I have my own copy that I edited, minus photographs, etc.

67 "Parliamentary Questions," 10 May 2007, www.europarl.europa.eu/sides/getDoc.do?type=WQ&reference=E-2007-2455&language=BG

68 See "Chemtrail Proof— German Military Exposed," www.youtube.com/watch?v=IaPqCMIuEk4

69 By "chaff," he means particulates. www.nogeoingegneria.com/news-eng/general-fabio-mini-chemtrails-in-the-sky-youre-convinced-and-so-am-i/

[Date unknown] 2009: While being interviewed at the European Parliament, Green Europarliamentarian Daniel Cohn-Bendit said, "The climatic dimension is a new dimension in which it is necessary to introduce the argumentation for disarmament."[70]

December 7–18, 2009: UN Copenhagen Climate Summit just after "Climategate" e-mails were "leaked." A whitewash.

After the Haiti earthquake on January 12, 2010, Claudia von Werlhof, professor of political science and women's studies at University of Innsbruck, Austria, was raked over the coals in the *Austrian Independent* after speaking out:

> Innsbruck political scientist Claudia von Werlhof has accused the USA of being behind the Haitian earthquake in January, it emerged today.
>
> According to a report on tirol.orf.at, Werlhof said that machines at a military research centre in Alaska used to detect deposits of crude oil by causing artificial earthquakes might have been intentionally set off to cause the Haitian earthquake and enable the USA to send 10,000 soldiers into the country.
>
> Ferdinand Karlhofer, the head of the Innsbruck Political Science Institute where von Werlhof works, has slammed her comments. He said such conspiracy theory had no scientific basis and her claim would damage the reputation of the Institute abroad.[71]

May 28–30, 2010: Chemtrails First International Symposium, Ghent, Belgium. Hosted by the Belgian environmental watchdog the Belfort Group, and attended by chemtrail awareness groups from Greece, Germany, Holland, France. The Belfort Group CitizensInAction works to raise public awareness of toxic aerial spraying.

U.S. American filmmaker Michael Murphy—*Environmental Deception, What in the World Are They Spraying?* (2010), and *Why in the World Are They Spraying?* (2012)—presented, and aerospace engineer Dr. Coen Vermeeren of the Delft University of Technology introduced the 300-page scientific report "CASE ORANGE: Contrail Science, Its Impact on Climate and Weather Manipulation Programs Conducted by the United States and Its Allies." Case Orange was prepared for the Belfort Group by a team of anonymous scientists, then sent to embassies, news organizations and interested groups around the world 'to force public debate.'[72]

---

70 See YouTube "Enouranois II: Persistent aircraft emissions — geoengineering — HAARP: Political dimensions," uploaded by the CSE Initiative on October 15, 2012.

71 "U.S. military behind Haiti quake, says Innsbruck scientist." *Austrian Independent,* 9 March 2010.

72 Rady Ananda, "Atmospheric Geoengineering: Weather Manipulation, Contrails and Chemtrails." *Global Research,* July 30, 2010.

July 4, 2010: Ecological Environmental Movement of Cyprus Green Party demonstration at the British Akrotiri Bases in Cyprus[73]

October 2010: Tenth Conference of Parties to the Convention on Biological Diversity in Nagoya, Japan:

> In a landmark consensus decision, the 193-member UN Convention on Biological Diversity (CBD) will close its tenth biennial meeting with a five-year de facto moratorium on geoengineering projects and experiments (like Solar Radiation Management), including no aerial spraying. "Any private or public experimentation or adventurism intended to manipulate the planetary thermostat will be in violation of this carefully crafted UN consensus," stated Silvia Ribeiro, Latin American Director of ETC Group.[74] Needless to say, the U.S. did not agree to the moratorium and has ignored it in the name of "research."

April 6, 2011: "Cypriot Greens Call for Chemtrail Investigation," *CyprusMail.*

> "We demand the government keep its promise and examine the possible consequences to the health of residents and the environment from chemtrails created by aircraft taking off from the British Bases in combination with the experimental operation of the High Frequency Active Auroral Research Programme (HAARP) within the Akrotiri British Bases and in Cyprus airspace in general," party head Ioanna Panayiotou said . . .The Greens said the government made a commitment to the House Environment Committee in March 2009 to conduct an in-depth study of these accusations. A bi-ministerial technical committee was subsequently formed for this purpose. The technical committee, the Greens added, "dissolved mysteriously" after only a few sessions. . .

On October 2, 2011, Nicola Aleksic, director of the Ecological Movement of Novi Sad, was unlawfully arrested for peacefully protesting against GMOs and chemtrails in front of the National Assembly of Serbia in Belgrade. Citizens of Serbia strongly support his efforts to defend the constitution and laws of Serbia against criminal aerosol spraying and GMO legalization. Listen to a translation of his impassioned speech to his people at "Against GMOs? Against chemtrails? Check this out." (Several YouTube clips of the speech are still up and running.)

On December 8, 2011, across the Atlantic Ocean in Suffolk County, New York, one hundred Americans showed up for a public hearing on banning geoengineering[75] from their skies, much as Aigina hoped to do.

---

73 Giannis Ioannou, "Aerial Spraying: Mobilization Around the Cyprus Bases." *Simerini*, June 25, 2010.

74 "Geoengineering Moratorium at UN Ministerial in Japan: Risky Climate Techno-fixes Blocked." News release, www.etcgroup.org, 29 October 2010.

75 Video footage at www.geoengineeringwatch.org/suffolk-county-public-hearing-to-ban-geoengineering/

July-December 2012: Cyprus Presidency of the Council of the European Union.

Permilla Hagberg, a parliamentarian leader of the Green Party in Vingaker, Sweden, ignited a media blitz after commenting on a photograph of chemtrails mislabeled as jet condensation ran in the Katrineholm *Kuriren* on 20 September 2012. Speaking as a private individual and not for the Green Party, Hagberg told the daily *Aftonbladet*:

> It is one of the most serious phenomena in Sweden today. These trails contain a multitude of chemicals, viruses, and heavy metals such as Aluminum that influence the weather. We have to convince the technicians of the airplanes to stop installing the tank containers. [Translated from Swedish]

Hagberg believes that the CIA and NSA are behind the program. "To be able to control the weather and use it for their own purposes is very advantageous for power structures. The Swedish government could also be involved." A Swedish meteorologist scoffed back that the trails in the sky are contrails, and another Green Party leader claimed that Hagberg was "inventing foolish things."[76]

January 12, 2013: Interview with Dimitris Kazakis—economist, general-secretary of the United People's Front (EPAM), and involved with ATTAC (Association for the Taxation of financial Transactions and Aid to Citizens)—on the Greek television interview show *Speaking Frankly* with host Konstantinos Bogdanos. "The forces of geoengineering that can destroy the planet are basically economic." (See Chapter 6, "Climate Engineering, Food, and Weather Derivatives.")

April 8–9, 2013: "Beyond Theories of Weather Modification: Civil Society Versus Geoengineering" conference hosted by the European Parliament in Brussels. Organizations under the "Skyguards" umbrella in collaboration with the Alternative-Political Laboratory and parliamentary groups the Greens / European Free Alliance and Alliance of Liberals and Democrats for Europe. To resume the work commenced in 1998 by the Committee on Foreign Affairs, Security and Defense Policy of the European Parliament, especially regarding the ignored proposal for a resolution "On the Environment, Security and Foreign Policy" (Europarliamentarian Maj Britt Theorin[77], 14 January 1999).

This 2013 European conference came full circle from Nick Begich's original testimony 15 years ago before the Committee on Foreign Affairs, Security and Defense Policy of the European Parliament. As an email overview of the conference from Wayne Hall concluded:

---

76 R. Teichmann, "Chemtrails, 'The Spraying of Poison over Sweden': Politician Ignites Controversial Debate." *Global Research*, October 20, 2012.

77 Theorin served in the Parliament of Sweden 1971–1995 and the European Parliament 1995–2004. Theorin's report instigated the February 1998 European Parliament public hearings on HAARP.

> The organizers demanded that the motion of the 14th January 1999 be implemented, given the abundance of documentary evidence now existing that underscores the seriousness of the [spraying] question. The European Parliament should establish the legal instrument of a parliamentary committee of enquiry without further delay. Some of the speakers will attempt to draw European public attention to this violation of European principles, assault on the environment and the health of citizens, and silence of complicity in concealing extremely serious wrongdoing that must end in legal and penal responsibility.[78]

The silence of complicity on this side of the Atlantic Ocean, where this blight upon the Earth originated, is deafening. It is high time we make a deep study of just how this one-two punch of HAARP-chemtrails works.

78 Email from Wayne Hall, April 11, 2013.

China HAARP Array | Chapters 1
43°04'51.75"N/92°48'26.85"E, near the villages of Jiefang and Kan'erjing, Google Earth gives us an aerial shot of what might be China's next generation of HAARP arrays.

Beale RADAR4 Facility | **Chapter 1**
The U.S. Air Force maintains five PAVE Phased Array Warning System (PAWS) Early Warning Radars (EWR). These radars are capable of detecting ballistic missile attacks and conducting general space surveillance and satellite tracking. The acronym PAVE is a military program identification code.

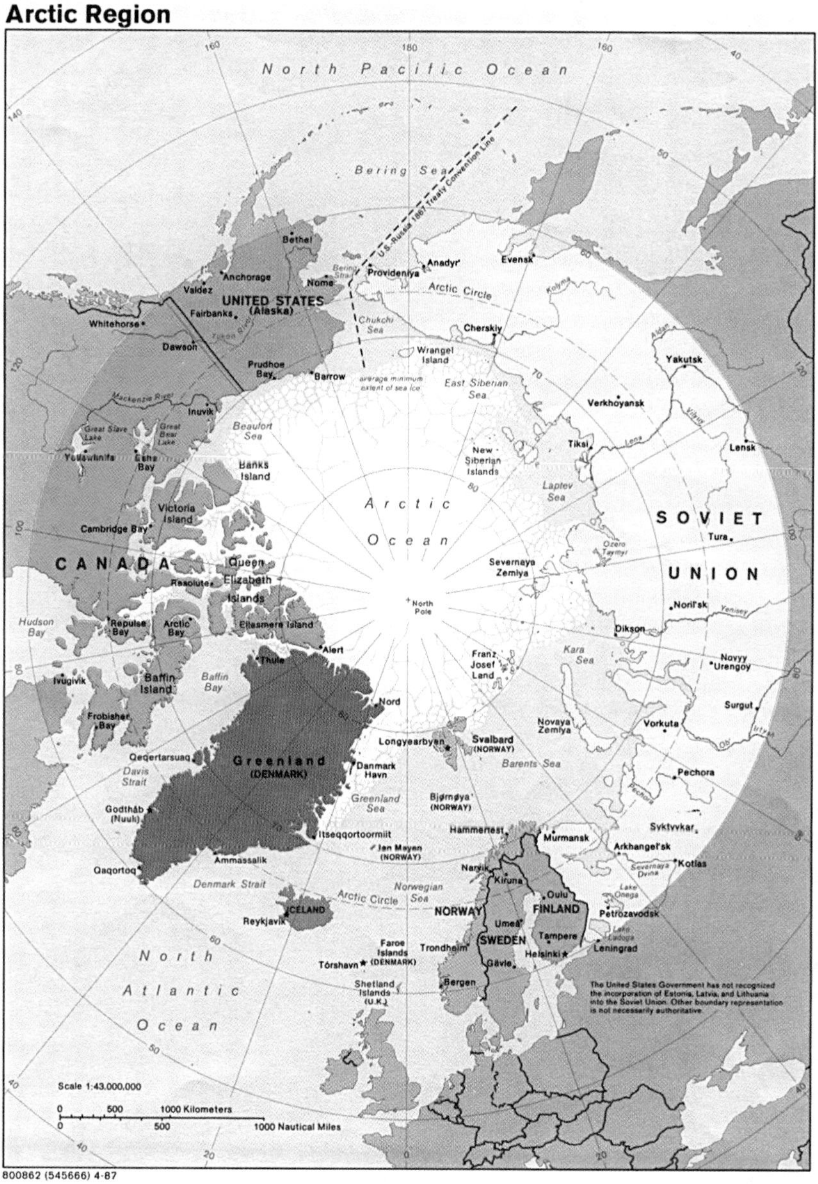

Arctic Circle Map | Chapter 1

Map of the Arctic Circle, the perfect latitude for Radio Frequency Radiation experiments.

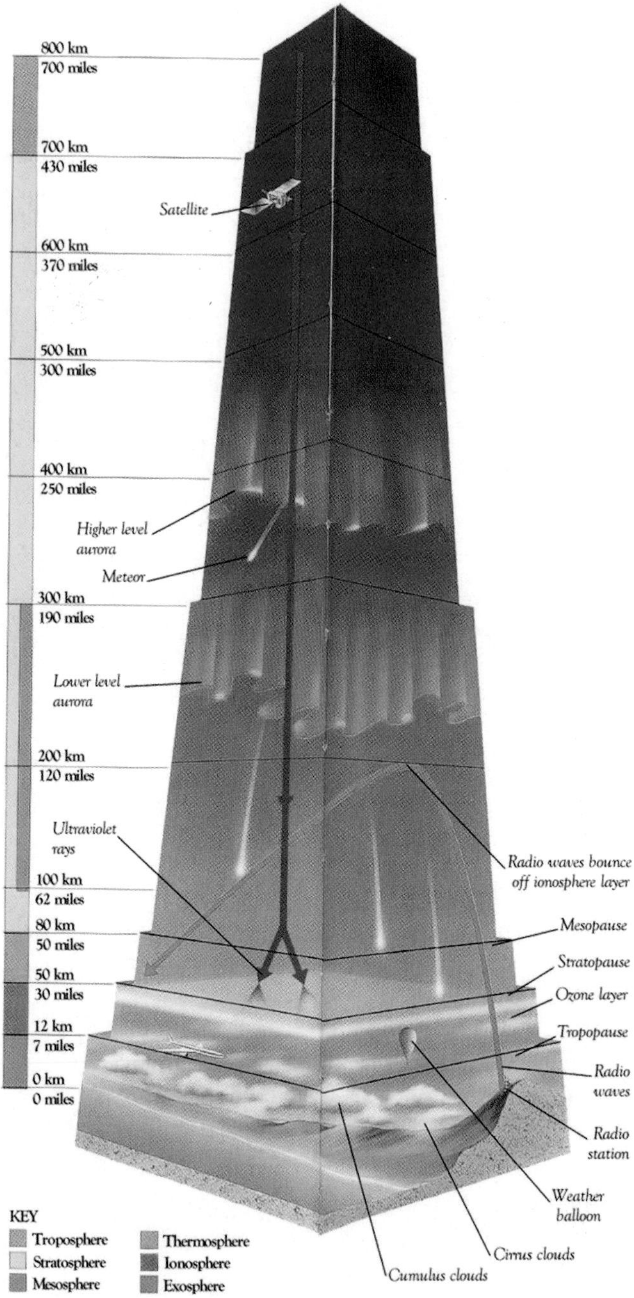

Layers of the Upper Atmosphere | Chapter 1

Regions of the Earth's atmosphere.

HAARP Array in Gakona, Alaska | **Chapters 2 and 7**
The first phase of HAARP construction was the Ionospheric Research Instrument (IRI), the high-power radio frequency transmitter that you can readily see in any photograph of the Gakona installation. On June 14, 1994, the Secretary of Defense requested an increase in HAARP funding purportedly to support HAARP's EPT (earth-penetrating tomography) and HAARP's VHF-UHF 3GHz to create artificial ionospheric mirrors for over-the-horizon radar.

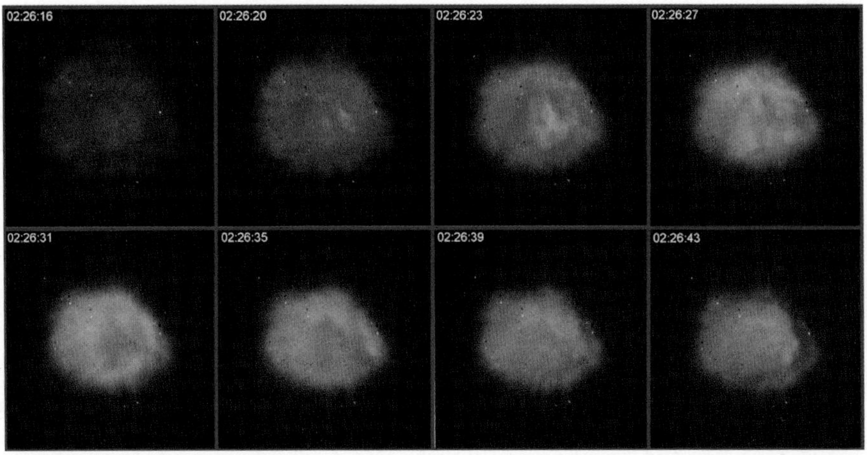

Artificial Plasma Cloud | **Chapter 2**

The interferometer creates scalar EM waves (Tesla waves, electrogravitational waves, longitudinal EM waves, waves of pure potential, electrostatic/magnetostatic waves, and zero vector waves). First, it sends out plasma orbs as target marker beacons—what people mistake for alien presences or ball lightning. Once the target is marked, the longitudinal wave is ready to bypass the 3-space world and instantaneously strike.

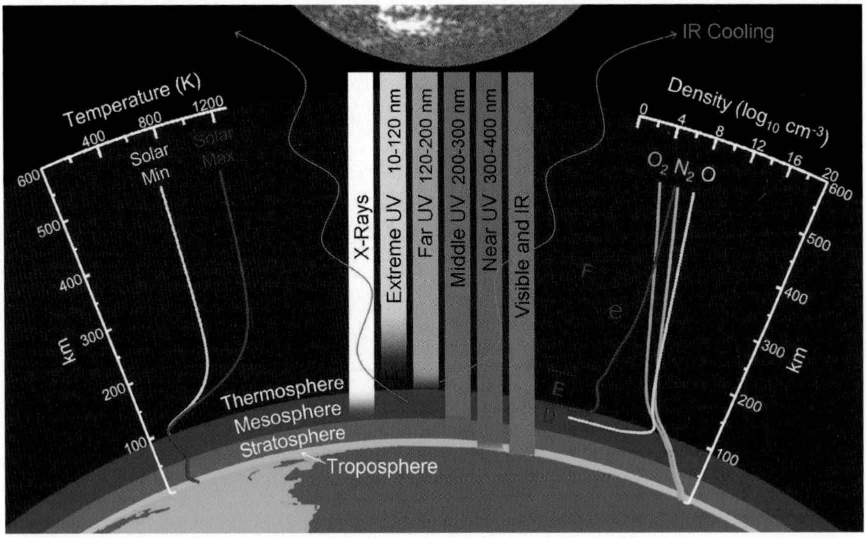

Earth's Upper Atmosphere | **Chapter 2**

Weather modification is possible by, for example, altering upper atmosphere wind patterns by constructing one or more plumes of atmospheric particles, which will act as a lens or focusing device. As far back as 1958, the chief White House adviser on weather modification, Capt. Howard T. Orville, said the U.S. defense department was studying "ways to manipulate the charges of the Earth and sky and so affect the weather by using an electronic beam to ionize or de-ionize the atmosphere over a given area" . . .

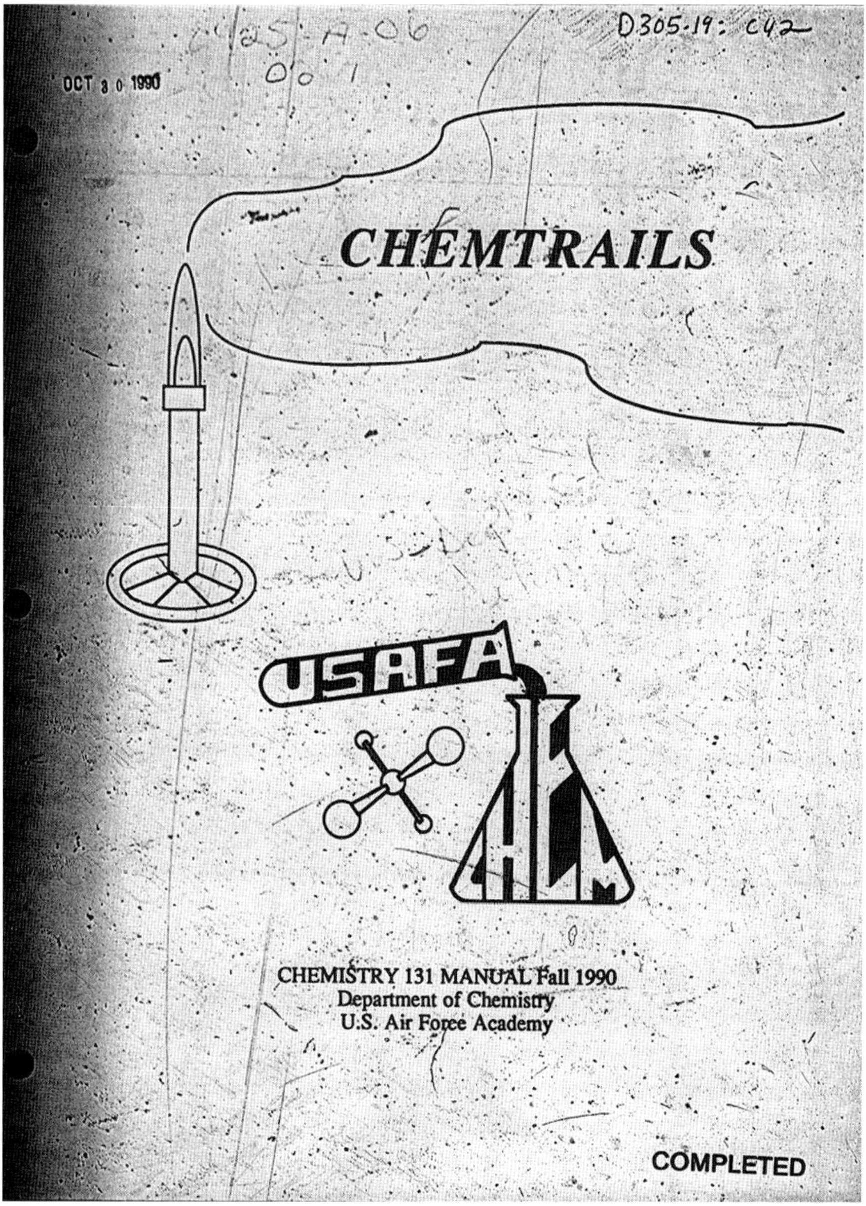

Chemtrails Manual | Chapter 3

The disparaged term "chemtrails" is a contraction for an aviation smog of "chemical trails" and hearkens from the 1990 U.S. Air Force Academy Chemistry 131 manual preparing cadets in molecular geometry, acid rain, spectroscopy, acid base titration, the chemistry of photography, identification of chemical compounds, chemical kinetics, electrochemistry, and organic chemistry in preparation for Air Force aerosol programs.

Circular Cloud Trail | Chapter 3
A circular "persistent contrail".

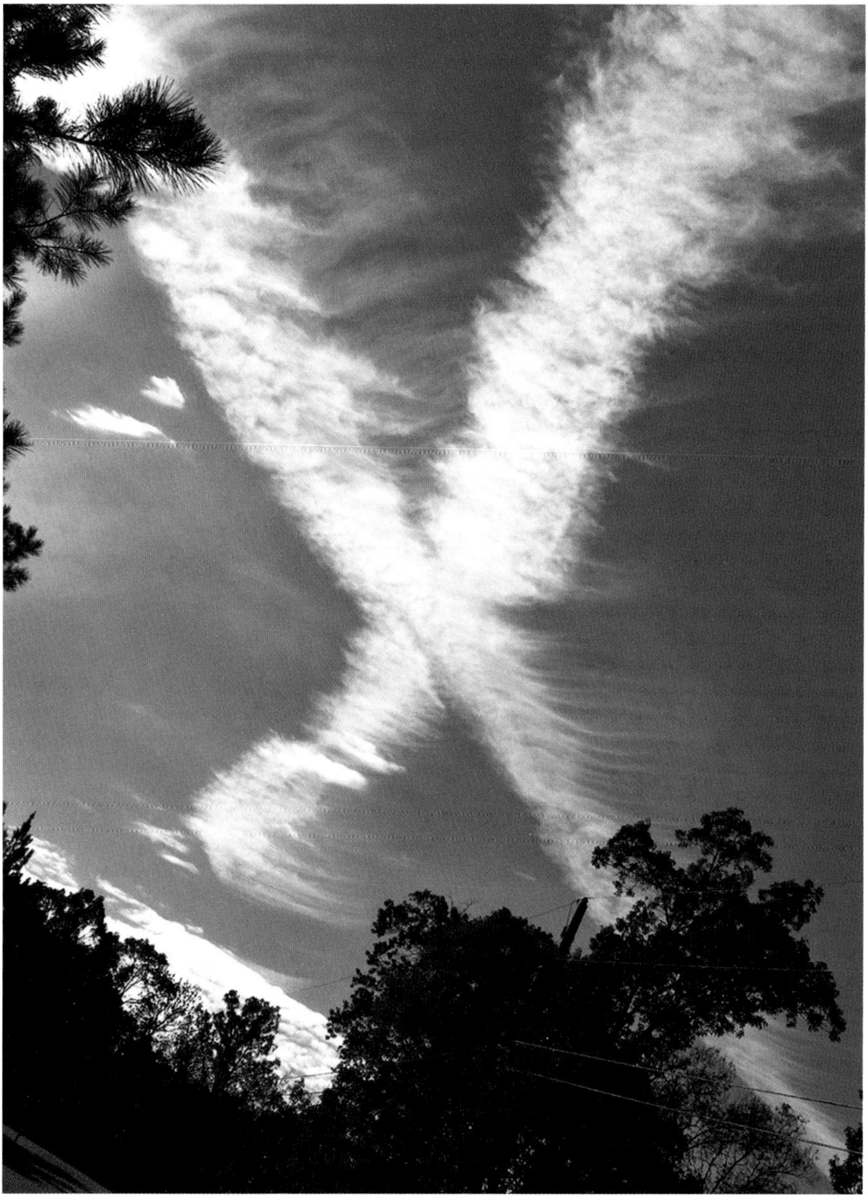

X-Cross Chem-Con-Trail | **Chapter 3**
How can it be that the trails laid out like crossword puzzles or strange circular forms spreading out into an overarching pale haze are the contrails and fumes of normal jet traffic?

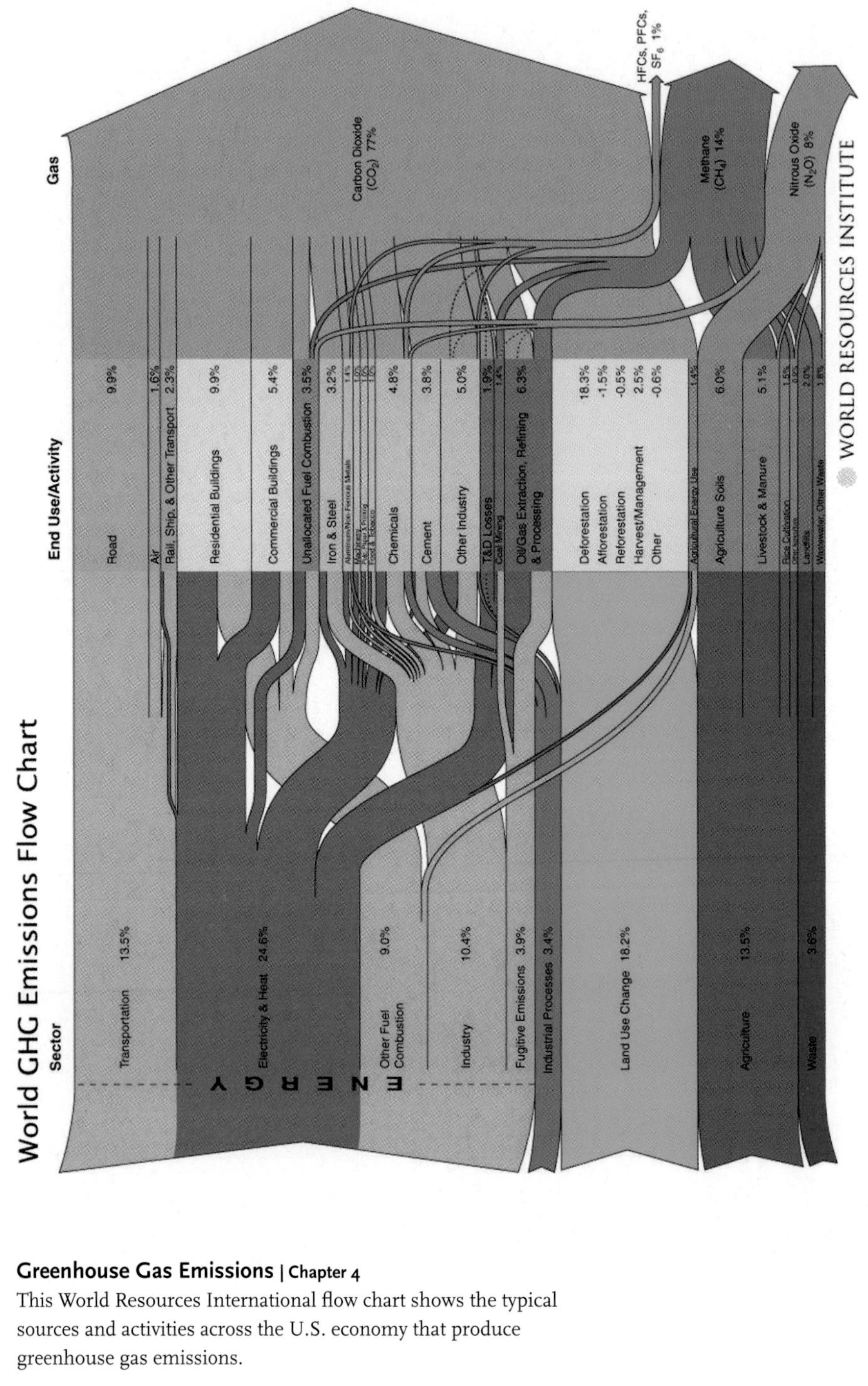

Greenhouse Gas Emissions | Chapter 4

This World Resources International flow chart shows the typical sources and activities across the U.S. economy that produce greenhouse gas emissions.

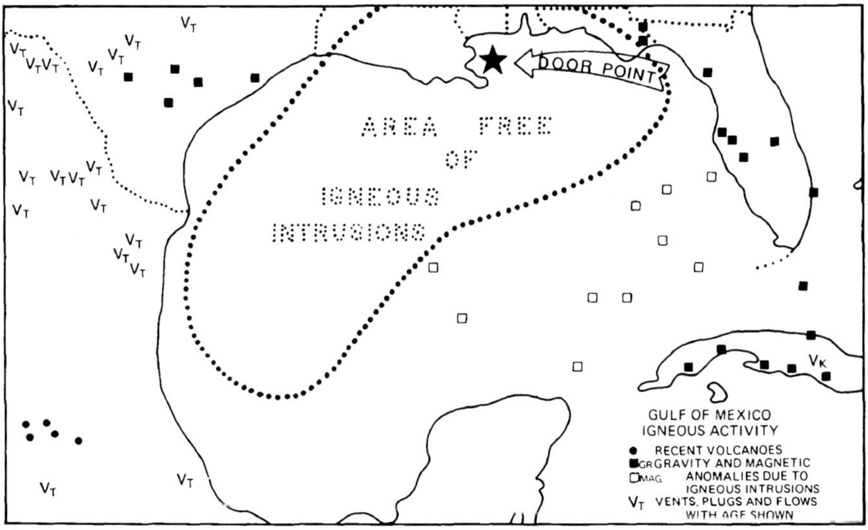

Door Point Volcano Map | Chapter 5
Both sinkholes and earthquakes are rising and sinking all over the country. It was after Hurricane Katrina in 2005 and Deepwater Horizon in 2010 that a 14-acre sinkhole opened up southeast of Baton Rouge. The cause? The Napoleonville salt dome being "farmed" by Texas Brine Company LLC by injecting millions of gallons of water had collapsed.

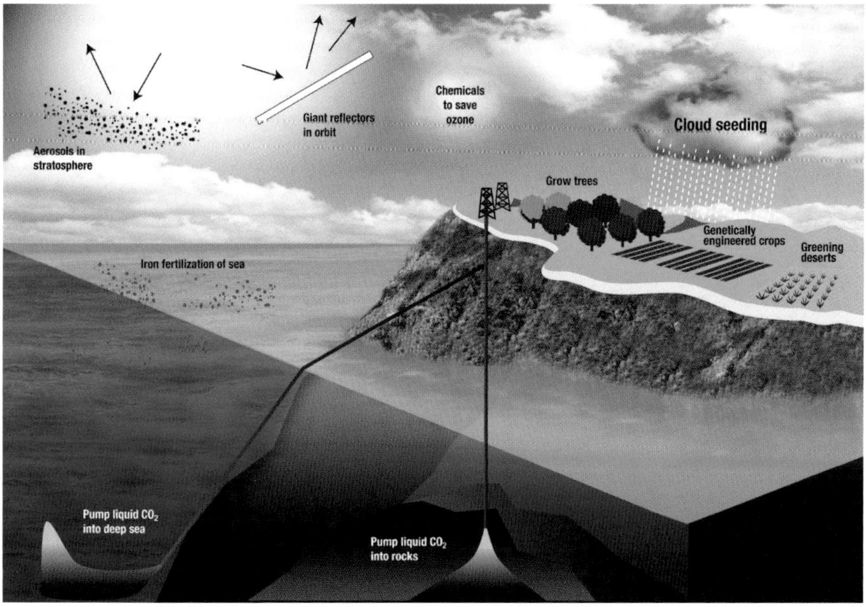

Geoengineering Basics | Chapter 6
The "positive spin" of Geoengineering. This illustration shows the basic concepts of altering the atmosphere to create new weather patterns.

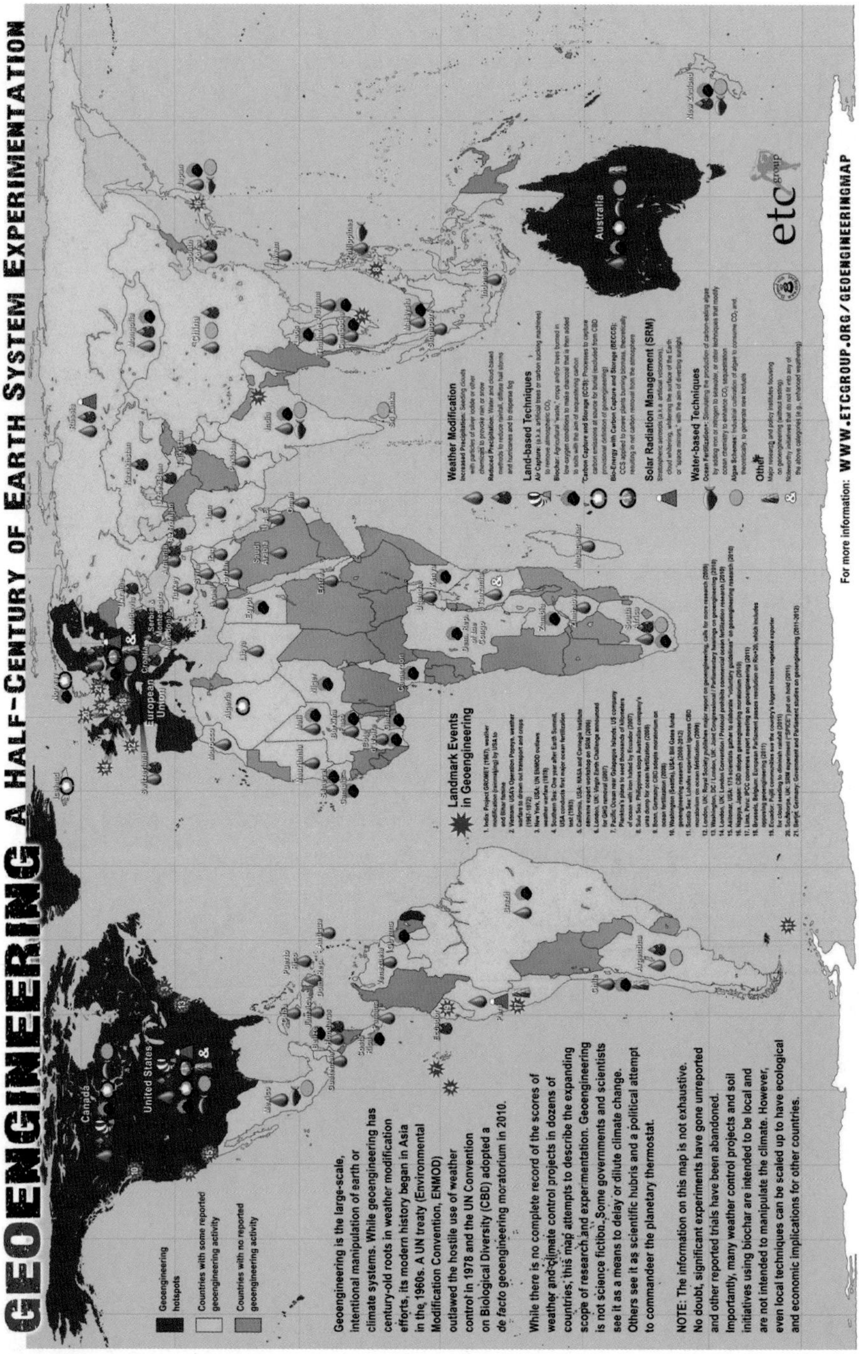

Geoengineering Map | **Chapter 7**

Map showing regions of geoengineering activity past and present around the world.

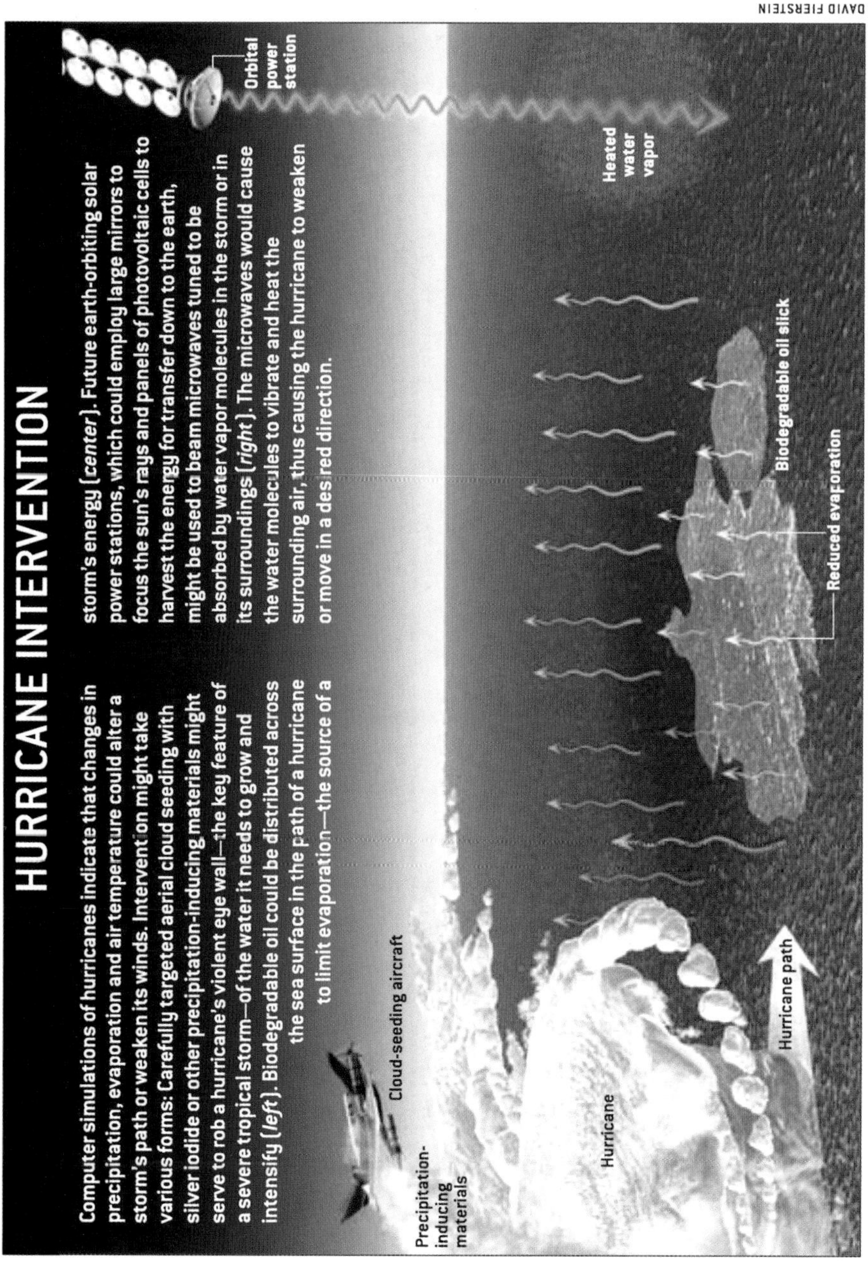

Hurricane Intervention | **Chapter 7**
The Russian corporation Elate Intelligent Technologies and its "Weather Made to Order" director Igor Pirogoff promise that for US$200 a day, Elate hurricanes can be steered within a 200-square-mile range, and that even Hurricane Andrew could have been decreased to a "wimpy little squall."

SBX in Seattle Harbor | Chapter 7
The SBX is a combination of the worlds largest phased array X-band radar carried aboard a mobile, ocean-going semi-submersible oil platform.

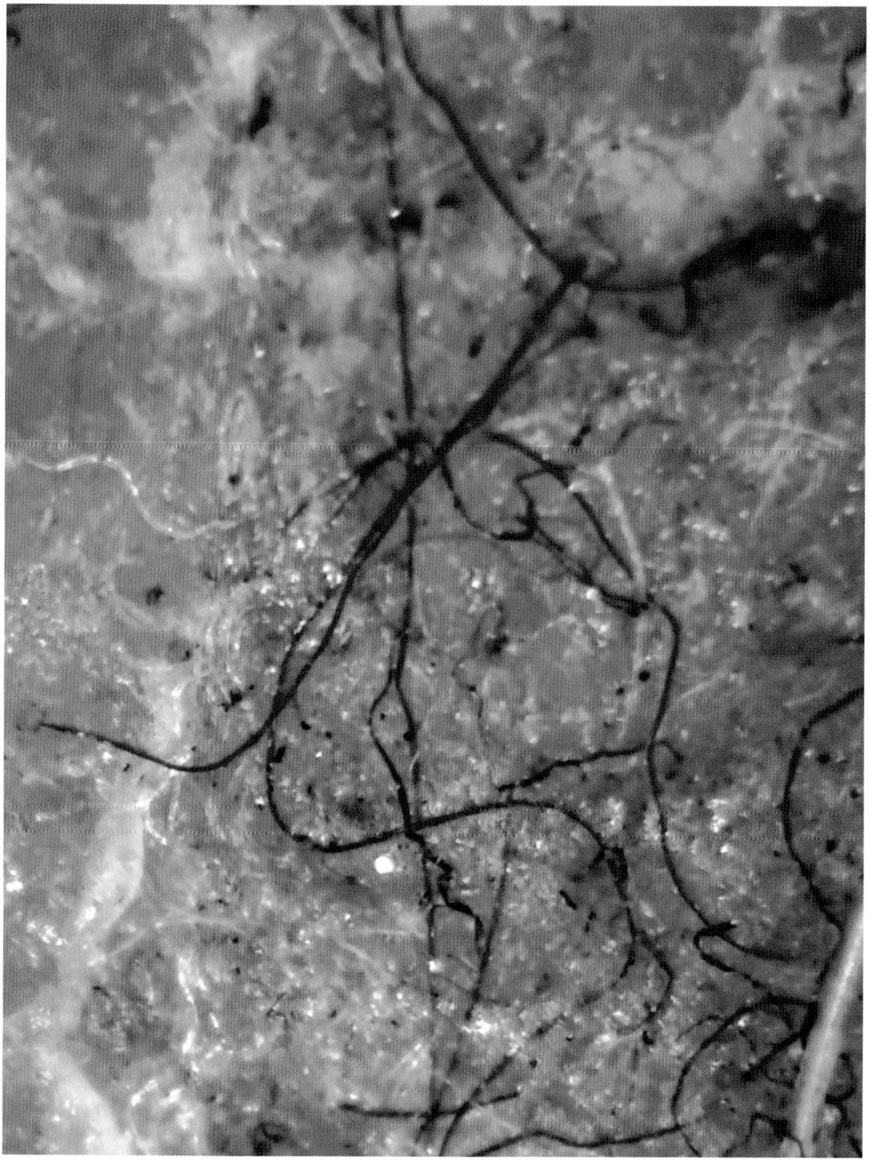

Morgellons Fibers | Chapter 8

Carnicom has successfully and repeatedly cultured the Morgellons pathogen. The significance of being able to culture a pathogen means it might be possible to control, inhibit, reduce, or eliminate similar pathogenic forms in the body.

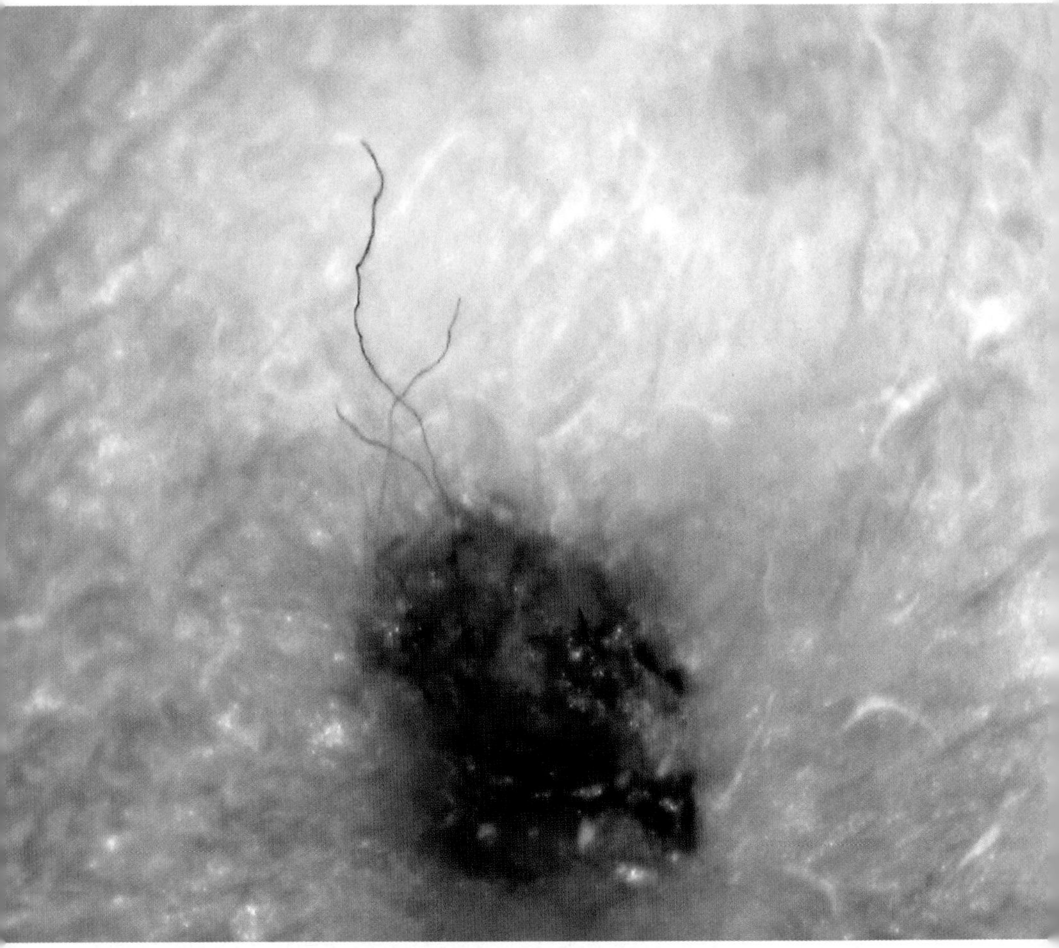

Morgellons Sore with Fibers | **Chapter 8**

The lesions can appear anywhere in a patient's body and quite often contain fiber-like strands or fibrous material. The fibers are the most perplexing visible feature of Morgellons. Often when an attempt is made to remove or extract the fibers the material will resist and act to withdraw or move away from whatever instrument is being employed.

CHAPTER FOUR

The Poisons Raining Down

I think the particles are to enhance their HAARP & other energy weapons for warfare, and to destroy our atmosphere so they can blame us for "global warming" to excuse imposing a global tax on us, robbing us of more of our wealth, and of course depopulation . . .

— Authentica, August 19, 2012

A method has been developed to isolate and record the existence of certain sub-micron particulates that appear to be resident within the atmosphere. The evidence from this research continues to support the claim of high levels of extremely fine metallic salts within the atmosphere as a consequence of the aerosol operations. The method developed incorporates a combination of ionization collection, electrolysis for separation, and significant advances in microscopy that have been made with relatively modest means. A significant chemical reaction demonstrating the presence of metallic ions from the atmospheric sample is presented.

— Clifford E. Carnicom, "Sub-Micron Particulates Isolated," April 26, 2004

The little bubble of air we breathe, crowded with transmission lines and cell phone (microwave) towers, is being dosed with jet fuel, silver iodide, and particulates—and that's not counting what's in the chemtrails that high-flying military jets and commercial airliners are laying. First, let's examine the jet fuel now considered a necessary evil of modern life.

When gauging how much JP-4 dosed with who knows what is actually affecting the unsuspecting below, remember to take into account that it is not unusual for tankers to dump thousands of gallons of fuel *in flight* in order to trim out the aircraft.

In the ninth edition of *Conceptual Physics*[1], Paul Hewitt points out that jet fuel itself creates plasma:

> Plasma Power: A higher temperature plasma is the exhaust of a jet engine. It is a weakly ionized plasma, but when small amounts of potassium or cesium

1 Paul Hewitt, *Conceptual Physics*. First published in 1971. Perfect book for the layman, plus the website www.physicsplace.com—a text-media combination for further readers' comprehension.

> metal are added, it becomes a very good conductor, and when it is directed into a magnet, electricity is generated. This is MHD power, the magnetohydrodynamic interaction between a plasma and a magnetic field.

Freelance journalist Jim Stone ("Truth is reality, which lies and inventions fall to in the end") actually equates jet fuel with the entire chemtrails delivery system:

> Chemtrails are not a major topic of mine, but it is important to know that though they are real, they are not actually being sprayed anymore per se, they have been mandated as a fuel additive and now all jet aircraft deliver them whether they want to or not. This transition happened about 8 years ago, when the grid patterns we once all saw suddenly got replaced by random hash left behind by commercial flights. That is why they look random now but have the same effect. It is also why they are not only over cities anymore, because they can't be turned off if they are just part of the fuel.[2]

Though it is absolutely probable that jet fuel is being treated with all manner of chemicals, that it constitutes the entire chemtrails program is not, the drawback being that even thousands of flights a day spewing jet fuel and exhaust cannot possibly produce enough plasma for continuous military-intelligence C4 operations and "full spectrum dominance." Thus the necessity for conductive metal particulates that can be further ionized and "electrified."

The Hughes Aircraft's 1991 Stratospheric Welsbach seeding patent points to the necessity of adding conductive particulates, despite the side effect of increased "global warming":

> One technique proposed to seed the metallic particles was to add the tiny particles to the fuel of jet airliners, so that the particles would be emitted from the jet engine exhaust while the airliner was at its cruising altitude. While this method would increase the reflection of visible light incident from space, the metallic particles would trap the long wavelength blackbody radiation released from the earth. This could result in net increase in global warming.[3]

As to what jet fuel might be spewing into our atmosphere, consider what is being loosed in the jet cabin. The cause of death of two 43-year-old British Airways (BA) pilots, Richard Westgate and Karen Lysakowska, point to the unfiltered air entering the cabin via a bleed pipe off the jet engine—fumes so strong that pilots often don oxygen masks. Westgate's lawyer, a pilot himself, is committed to suing BA and proving the existence of *aerotoxic syndrome*, a chronic physical and neurological condition that thousands of commercial

2 Email from Jim Stone, 12/5/12. See www.jimstonefreelance.com/haarp.html

3 "Background of the Invention," www.google.com/patents/US5003186

pilots suffer from. BA is liable under the Control of Substances Hazardous to Health regulations for failing to monitor air quality on board its planes.[4]

Jet fuel toxicity is practically a synonym for *organophosphates*—not to mention the depleted uranium (DU) counterweights ranging in weight from 0.23 kg to 77 kg still being used on the McDonnell Douglas MD-II and DC-10s, Boeing 747s, Lockheed C-130s, C-141s, Jetstars, and S-3As.

Westgate complained of persistent headaches, memory loss, chronic fatigue, mood swings, and an inability to multitask (deadly when flying a plane)—all typical of the aerotoxic syndrome many complain of after undergoing "chembombing" below. Flight attendants are filing formal complaints about itchy skin, lesions, hair loss, and loss of memory. The tributyl phosphate in their uniforms is being blamed,[5] but are the organophosphates they're breathing the real cause?

Westgate was misdiagnosed with depression, just as "chembomb" and Morgellons sufferers are misdiagnosed. In an attempt to look a little deeper, specialist aviation medic Dr. Michel Mulder observed, "Some of the symptoms are like the early onset of Parkinson's disease or MS. There needs to be an understanding of this but it's willfully not recognized."[6] Parkinson's and multiple sclerosis are *neurological* diseases.

The Hertfordshire-based autonomic neurophysiologist Dr. Peter Julu—Europe's *only* autonomic neurophysiologist—believes it is the lack of experts in his field that prevents full recognition of organophosphate poisoning from toxic fumes, whether it's jet oil or sheep dips. (He has found similar symptoms among farmers.) Organophosphates attack the brain stem, which runs the autonomic nervous system, especially the part that deals with emotion and short-term memory. Dr. Julu is now treating many senior pilots.[7]

Organophosphate chemical nerve agents were directly implicated in the chronic multisymptom disorder known as the Gulf War Syndrome:

> Organophosphate poisoning results from exposure to organophosphates (OPs), which cause the inhibition of acetylcholinesterase (AChE), leading to the accumulation of acetylcholine (ACh) in the body. Organophosphate poisoning most commonly results from exposure to insecticides or nerve agents. OPs are one of the most common causes of poisoning worldwide, and are frequently intentionally used in suicides in agrarian areas. . .[8]

Organophosphates are just the beginning. Above the Earth's atmosphere, rocket exhaust, space shuttles, and space capsules are ejecting chemicals designed to interact with charged particles and artificially light up the aurora

4 Ted Jeory, "Dead BA pilots 'victims of toxic cabin fumes'." *Sunday Express*, January 27, 2013.
5 "Alaska Airlines flight attendants claim new uniforms make them ill." *NBCNews*, May 4, 2012.
6 Jeory, "Dead BA Pilots," January 27, 2013.
7 "Alaska Airlines flight attendants," May 4, 2012.
8 en.wikipedia.org/wiki/Organophosphate_poisoning#Gulf_War_Syndrome

borealis (and even create an artificial aurora). *The ionosphere is being injected and transformed.* Strategic Defense Initiative (SDI) weapons systems, whatever their project names, are producing unstable, highly charged particle "clouds" in the atmosphere and Van Allen radiation belts for space flight C4 systems[9]—all of which should give new meaning to the term *trickle-down effect.*

AS ABOVE, SO BELOW

What happens when nanosized endocrine-disrupting chemicals whose propensity is to synergistically interact with the billions of solid and liquid particulates are suspended in the atmospheric aerosols? Surely, they will combine viruses, bacteria, mold spores, and pollen with anthropogenic (man-made) polluters like gas molecules, soot, tobacco smoke, smog, oil smoke, fly ash, cement dust, and suspended atmospheric dust. And what of the nanosized titanium dioxide in nutritional supplements, toothpaste, gum, candy, Pop Tarts, coffee creamers, purified water, over-the-counter drugs, and sunscreen; the neonicotinoids in pesticides[10]; the lead in rice imported from Asia, Europe, and South America that is 30–60 times greater than the FDA's provisional total tolerable intake (PTTI)[11]; the antimicrobial silver particles in clothing? And that's not including the Heinz 57 drifting down from above, none of which are covered by the Food and Drug Administration (FDA).

So far, we have discussed the organophosphates in jet fuel. Now, let's break down what the chemtrails themselves are spewing into the atmosphere for various military-industrial agendas. Picture 55-gallon drums filled with liquid chemical brews unloaded from trucks at airports and loaded into jet disposal systems or filling their entire fuselage. The hoses attaching the barrels to wing nozzles are hooked up according to exact directives for specific mixes over specific geographic or urban areas. The jets roll out onto the military or private tarmac and take off, some with pilots, some flown remotely, even some with passengers breathing the cabin air. Air traffic controllers have been given a heads-up and will not log the flights. The jets climb to 25,000–35,000 feet (military jets sometimes fly as low as 16,000 feet), level out, and head for the target latitude-longitude, at which point the nozzles open and a fine spray is released. . .

On December 7, 2007 and again in February 2008, a "yellow cloud" exploded in the air over Cedar Glen in the San Bernardino mountains of California, releasing a fine, slippery yellow dust like silt. "Cedar pollen," the media called

9 Paul Schaefer, a Kansas City electrical engineer who claims to have built nuclear weapons for four years, is referenced by Begich and Manning, as is his book *Energy and Our Earth*, but I could find nothing more on him.

10 Drew Mikkelsen, "Beekeepers call for state to investigate spike in bee deaths," KING5.com, April 11, 2013.

11 Alexandra Sifferlin, "Worrisome Levels of Lead Found in Imported Rice." Time.com, April 11, 2013.

it, despite the fact that it was winter. Was the bright yellow *sulfur*? And still we're expected to believe that KC-135s and KC-10s stripped of military markings and outfitted with spraying devices have nothing to do with tic-tac-toe grids or "yellow clouds" exploding "cedar pollen"[12] or "bee defecation":

> A Harvard scientist said the Yellow Rain Hmong people experienced was nothing more than bee defecation. My uncle explained Hmong knowledge of the bees in the mountains of Laos, said we had harvested honey for centuries, and explained that the chemical attacks were strategic; they happened far away from established bee colonies, they happened where there were heavy concentrations of Hmong. Robert grew increasingly harsh, "Did you, with your own eyes, see the yellow powder fall from the airplanes?" My uncle said that there were planes flying all the time and bombs being dropped, day and night. Hmong people did not wait around to look up as bombs fell. We came out in the aftermath to survey the damage. He said what he saw, "Animals dying, yellow that could eat through leaves, grass, yellow that could kill people—the likes of which bee poop has never done."[13]

PARTICULATE MATTER (PM)
1 nanometer = 1 billionth of a meter
Coarse particle = 10,000–2,500 nanometers
Fine particle = 2,500–100 nanometers
Nanoparticle (ultrafine) = 1–100 nanometers
1 micron = 1/70 human hair

In 1960, the U.S. National Bureau of Standards adopted the prefix *nano-* for "billionth," and nanotechnology was officially born. Truly, the microscopic size of materials being dropped from the sky and entering our lungs, bloodstreams, and soil defies belief. The term *nanometer*, formerly known as millimicrometers or millimicrons (*m*☜), which is 1/1000 of a micron (☜), has entered the international language of science with a vengeance.

> Nanotechnology is an emerging field of science that deals with engineered molecules a few billionths of a meter in size. Because of the novel arrangements

12 Arthur Cristian, "Increased Illness Linked to Mystery Powder — More Mountain Patients Reporting Sickness Since the Yellow Dust Appeared." *The Alpenhorn News*, June 26, 2007.

13 Kao Kalia Yang, "The Science of Racism: Radiolab's Treatment of Hmong Experience." *Hyphen* magazine, October 22, 2012.

> of the atoms in these molecules—and because the laws of physics behave differently at such scales—nanoparticles display bizarre chemical properties. Those properties make them potentially useful in products including stain-proof fabrics and computer components, but also make them potentially biologically disruptive.[14]

Nanoparticles are incredibly small carbon molecules often engineered with tinier than you can imagine radio-controlled gigaflop microprocessors now called "smart" molecules or "smart dust": sensors made from mono-atomic gold particles linked to supercomputers the size of a grain of sand. Or as Brewer Science puts it, carbon nanotubes (CNTs) are "flexible electronics [that] would allow high-tech devices to be attached to things that naturally bend and stretch, such as clothing and even human tissue."[15]

Chemtrails are loaded with nanosensors. According to industrial toxicologist Hildegarde Staninger, Ph.D., when these sensors fall like bad fairy dust, they look like "iridescent glitter." In the literature, they are called MEMS (microelectromechanical sensors) and GEMS (global environmental MEMS sensors)[16]. They record everything, are inhaled and ingested everywhere, *and can be remotely communicated with.* After all, what good would sensors be if you couldn't access what they're sensing? Needless to say, smart dust research is heavily funded by the Department of Defense (DoD).

> Fitted with computing power, sensing equipment, wireless radios and long battery life, the smart dust would make observations and relay mountains of real-time data about people, cities and the natural environment.[17]

The human price of smart dust and nanoparticles raining down on all biological life has yet to be tallied, but we do not need rocket science nor plasma physics to tell us that it will be vast. In December 2006, the City of Berkeley, California amended its hazardous materials law to include nanoparticles, given that the Center of Integrated Nanomechanical Systems is on the UC Berkeley campus. In "Cornell sounds nanoparticle health warning"[18], Anne Ju discusses how polystyrene nanoparticles—polystyrene being a common, FDA-approved material found in consumables from food additives to vitamins—cause barely

14 Rick Weiss, "Nanoparticles Toxic in Aquatic Habitat, Study Finds." *Washington Post*, March 29, 2004.

15 "Brewer Science Brings Game-Changing Flexible Electronic Devices Closer to Reality," January 15, 2013, www.brewerscience.com

16 "Global Environmental MEMS Sensors (GEMS): A Revolutionary Observing System for the 21st Century," ENSCO, Inc., prepared for NASA Institute For Advanced Concepts, 2 December 2002.

17 John D. Sutter, "'Smart dust' aims to monitor everything." *CNN*, May 3, 2010. Note the use of the subjunctive mood.

18 *eats shoots 'n leaves,* a *Cornell Chronicle* commentary, richardbrenneman.wordpress.com/2012/02/17/cornell-sounds-nanoparticle-health-warning/

detectable changes but lead to an overabsorption of harmful compounds. An article at *nanotechwire* discusses how inhaling carbon black nanoparticles, like the ones in atmospheric aerosols that have been deployed for decades[19] to melt Arctic ice ("dark ice") and intensify hurricanes, kills macrophages—the immune cells in the lungs responsible for attacking infections—and doubles lung inflammation (pyroptosis). Particulates of all stripes, especially those of less than a ten-micron width, nestle into the deepest parts of lungs and damage their functions, leading to heart disease, premature aging, and death.

Did I say that the smart dust trail of tears is sure to be long?

Nanoparticles lodge in the soil just as they do in the body. A study of the earthworm *L. rubellus* by the Dutch research institute Alterra revealed one side effect of nanotube "flexible electronics": 154 milligrams of carbon nanoparticles in a kilogram (2.2 pounds) of soil produced 40 percent fewer cocoons and showed increased mortality and tissue damage. Doctoral candidate Merel van der Ploeg at Wageningen University in the Netherlands concluded:

> The same characteristics which make nanoparticles useful in many products, such as chemical reactivity and persistence, cause concern about their potential adverse health effects . . .Nanotechnology deals with matter on a scale comparable to the diameter of a strand of DNA, where *materials can work differently when it comes to things like chemical reactions and electrical conductivity.*[20] [Emphasis added.]

"Chemical reactions and electrical conductivity" are exactly what chemtrails and HAARP are all about.

A lawsuit against the Environmental Protection Agency (EPA) for having exposed humans to deadly inhalable particulates (PM2.5 from a diesel truck) at its Human Studies Facility at the University of North Carolina at Chapel Hill stated as evidence that: "2005 levels of PM2.5 and ozone were responsible for between 130,000 and 320,000 PM2.5-related and 4,700 ozone-related premature deaths."[21] EPA administrator Lisa Jackson told the Oversight and Investigations Subcommittee of the House Energy and Commerce Committee, "Particulate matter causes death. It doesn't make you sick. It's directly causal to dying sooner than you should." The EPA is now barred from conducting life-threatening experiments, but will the injunction have any teeth?

For filtering aerosol particulates, Carnicom recommends the HEPA air filter (down to 0.3 microns), Ionic Breeze Electrostatic Air Cleaners (Model # 51637), and distilling rainwater via crystalline chemistry.

---

19 William M. Gray, "Feasibility of Beneficial Hurricane Modification by Carbon Black Seeding." Dept. of Atmospheric Science, Colorado State University, Fort Collins, CO, 1975.

20 Rudy Ruitenberg, "Earthworm Health Hurt by Nanoparticles in Soil in Alterra Study." *Bloomberg*, January 29, 2013.

21 Paul Chesser, "EPA Sued Over Heinous Experiments on Humans." National Legal and Policy Center, September 25, 2012.

POLYMERS

Polymer *goo* fell on Oakville, Washington in 1994, similar to the goo that turned icicles greenish yellow in Snyder, New York in 2011[22], and the goo near the Inupiat Eskimo village of Kivalina that looked remarkably like salmonberries[23]. Were they Dyn-O-Mat-like tests? Since 2001, Dyn-O-Mat has created Dyn-O-Storm and Dyn-O-Gel, making one wonder why the EPA even bothered with this warning:

> There are three categories or types of High Molecular Weight (HMW, >10,000 daltons) polymers typically reviewed by the New Chemicals Program: soluble, insoluble/non-water absorbing ("non-swellable"), and water absorbing ("swellable"). EPA has a concern for potential *fibrosis of the lung or other pulmonary effects that may be caused by inhalation of respirable particles of water-insoluble HMW polymers.* The toxicity may be a result of "overloading" the clearance mechanisms of the lung. EPA also has concerns for water absorbing polymers, based on data showing that cancer was observed in a 2-year inhalation study in rats on a HMW water-absorbing polyacrylate polymer.
>
> . . .For [water-absorbing (swellable) polymers], the Agency makes the "may present an unreasonable risk" determination with concerns for fibrosis and cancer, based upon water absorption properties. Concerns are associated with substances that absorb their weight (or greater) in water. The primary reference for Agency concerns for this class of polymers is TSCA 8(e)-1795, submitted by the Institute for Polyacrylate Absorbents (IPA), which indicated that high molecular weight polyacrylate polymers caused lung neoplasms in animal studies.[24] [Emphasis added.]

"Spider web"-like polymers (plastics) often seen on the ground and in trees after chemtrail tracks are laid are tiny synthetic filaments whose ingredients are kept separate in liquid form in two canisters on KC-135 Stratotankers, KC-10 Extenders, or commercial aircraft until the moment they are to be sprayed. At that point, the contents of the two canisters are combined to become the spun polymer threads to which nanoparticulates, pathogens, and mycoplasma will either cling or be encased in, as in the case of the polyethylene (HDPE) that encases Morgellons pathogens (see Chapter 8):

> Cases of people with Morgellon's [sic] Disease are increasing at a rate of 1,000 victims per day. In 2008, the CDC (Centers for Disease Control) began a study on Morgellon's to investigate its causes and symptoms. With Morgellon's,

22 Olivia Katrandjian, "Mystery Goo Turns Icicles Green and Yellow in Snyder, N.Y." *ABC News*, January 20, 2011.

23 Rachel D'Oro, "Orange goo near remote Alaska village ID'd as eggs." AP, August 9, 2011.

24 "High Molecular Weight Polymers in the New Chemicals Program," www.epa.gov/oppt/newchems/pubs/hmwtpoly.htm

> individuals exhibit unhealing sores containing fibers that burn at 1700°F and do not melt. A private study to determine the chemical and biological composition of these self-replicating fibers has shown that the fibers' outer casing is made up of high-density polyethylene fiber (HDPE). This material is used throughout the bio-nanotechnology world as a compound to encapsulate a viral protein envelope with DNA, RNA, etc.[25]

By mimicking a newborn spider's "parachute" filament, polymer tufts can lengthen the time and distance of nanoparticles in the upper atmosphere. On September 20, 2000, Carnicom issued "Project Report No. 1" regarding evidence of polymer fibers after aerosol discharges by subsonic aircraft. He notes that electroactive polymer fibers are identified and described in Defense Advanced Research Projects Agency (DARPA) documents[26]. Carnicom points to several polymer applications, especially when combined with the conductive metal barium:

> (1) Aircraft cloaking when irradiated;
> (2) Advanced RADAR applications;
> (3) Biological applications, including delivery and detection;
> (4) C4 (command, control, communications, computers) applications, including VTRPE (Variable Terrain Radio Parabolic Equation) computer propagation; and
> (5) Weather modification.
>
> In 1990, NATO detailed how sprayed polymers absorb *electromagnetic* radiation, not solar radiation,[27] which may have something to do with mixing polymer fibers with barium salts to "cloak" aircraft from radar (and eyes).

Chemical brain injury can occur when the plastic and metal particulates raining down on us couple with the industrial chemicals we are already ingesting and breathing. In 2009, Dr. Staninger reported that exposure to aerosols filled with nanocomposites inhibits the cholinesterase that the brain, liver, and red blood cells need. Chronic inhibition of this enzyme causes slow-death poisoning leading to neurological disorders and paralysis.[28] Dementia and anemia have been linked.[29]

---

25 "Crime Syndicate: Banksters, 9/11, BP, Chemtrails, Monsanto." *GeoEngineeringExposed,* December 2, 2012.

26 DARPA is an original funder of HAARP, along with the U.S. Air Force, U.S. Navy, University of Alaska, BAE Advanced Technologies, and Raytheon.

27 H. Jeske, Hamburg University, "Modification of Tropospheric Propagation Conditions," October 1990.

28 Hildegarde Staninger, Ph.D., "Exposure to Aerosol Emissions of Nano Composite Materials Resulted in Cholinesterase Inhibition," September 7, 2009. 1cellonelight.com/pdf/NanoCompositeCholinesteraseInhibition10.2009.pdf

29 "Anemia Linked to Increased Risk of Dementia." *ScienceDaily*, July 31, 2013.

MOLD AND FUNGI

Due to static charges, polymers attract mold and fungi spores, so mold-growth suppressants are sometimes added to the polymer mix, too. The fungus *Aspergillus fumigatus* was sprayed throughout the 1950s—a contaminant producing lesions and infections of lungs, bronchi, ears, sinuses, orbital bones, meninges, and often ending in death. The racist assumptions in the following 1980 news release are noteworthy, given the history of nonconsensual populations chosen for many biological and chemical warfare "experiments":

> Within this [Naval Supply System at Norfolk Supply Center, Virginia], there are employed large numbers of laborers, including many Negroes, whose incapacities would seriously affect the operation of the supply system. Since Negroes are more susceptible to coccidioides than are whites, this fungus disease was simulated by using Aspergillus fumigatus Mutant C-2.[30]

Bacillus subtilis spores share characteristics with *Bacillus anthracis.* "The secret spraying of *Bacillus subtilis* and other agents over populated areas had long been underway and would continue for years. Millions of subjects at all stages of life and in all manner of infirmity were being exposed. Moreover, exposed individuals were breathing not just hundreds of spores . . .but millions."[31]

On June 16, 2001, the CDC listed chronic lower respiratory disease as the fifth leading cause of death in the U.S. "from unknown airborne causes."[32] In early 2003 in the high dry New Mexico desert, Carnicom discovered 40+ colonies of molds in a condensation sample. This is important because molds are often closely related to respiratory, asthmatic, and allergic reactions. In July of the same year, polymer chemist R. Michael ("Mike") Castle, Ph.D., of Castle Concepts, Inc., identified microscopic polymers comprised of genetically engineered fungal forms mutated with viruses, warning that we are all breathing in *trillions* of *Fusarium*/virus-mutated spores that secrete a powerful mycotoxin.[33]

An increase in drought means an increase in the potentially lethal but often misdiagnosed *valley fever*. Accompanied by flu-like symptoms and severe (often bloody) coughs, the fever arises from breathing in *Coccidioides* fungus-laced spores in airborne dust that would certainly multiply in the drought conditions of a hotter, artificially ionized atmosphere, not to mention the possibility that the fungi might be loaded onto polymers for "experimental" purposes. Cases of valley fever rose by more than 850 percent between 1998 and 2011, the

---

30 News release from the Office of Congressman G. William Whitehurst, 2nd District Virginia, September 17, 1980. (Freedom of Information Act request)

31 Leonard A. Cole, *Clouds of Secrecy*. Rowland & Littlefield, 1988.

32 Clifford E. Carnicom, "Molds Flourish, Illness Prevails," May 10, 2003. This report merited a visit from the DOE's Joint Genome Institute associated with Lawrence Berkeley National Laboratory.

33 Dr. Mike Castle, "Chemtrails, Bio-Active Crystalline Cationic Polymers," July 14, 2003.

very years that the chemtrails program was quietly kicked into high gear. In California, there were 700 reported cases in 1998 and 5,500 in 2011; in Arizona, 1,400 in 1998 and 16,400 in 2011.[34]

With all of these chemically caused but misdiagnosed symptoms, what exactly *is* the flu now? On January 12, 2013, New York governor Andrew Cuomo was faced with 19,128 cases of flu as compared with 4,404 the entire 2011–2012 flu season. (In Boston, there were *ten times* as many cases as in 2012.) Cuomo declared a public health emergency and gave pharmacists permission to administer flu vaccinations to children as well as adults.[35] Was the epidemic more about worn-out immune systems after 20 years of chemtrails? Or was mycoplasma being dropped over New York City?

METALS

In *Chemtrails Confirmed,* Will Thomas tells how in 2000 airline pilot Captain D.A. Wheeler was puzzling over some high-altitude cirraform clouds that weren't drifting with the upper winds when he discovered *strong radar returns* immediately over them. At those altitudes, cirrus clouds are usually composed of ice crystals with no convective activity possible, due to low humidity. Captain Wheeler wondered if the strange echoes indicated the presence of a *metallic substance* to allow for better tracking of drift.

Interestingly, the Cambridge University Press publication in 2000, *Introduction to Atmospheric Chemistry* by Peter V. Hobbs, includes neither titanium, aluminum, barium, magnesium, nor calcium in a chart of expected components of the atmosphere.[36] A.K. Johnstone, Ph.D., begs to differ. He examines how a chemtrail mix of radar-sensitive chaff (loaded with barium and aluminum) and polymers (chemically synthesized compounds of large molecules)

> can be utilized as a dielectric capable of sustaining an electric field and transmitting electrostatic force. A dielectric and the surface of a metal, such as aluminum or barium, can maintain an electrostatic charge for a long period of time. The electrostatic force created is able to change the density of the air and, in turn, the air pressure.[37]

The polymers and metals being disseminated by daily chemtrail drops into our plasma-ized atmosphere are designed for multiple purposes, two being to disrupt radar signals and create corridors ("ducting") for HAARP's ionized

---

34 Gosia Wozniacka, "Fever hits thousands in parched West farm season." AP, May 6, 2013.

35 "New York governor declares public health emergency to combat flu." Reuters, January 12, 2013.

36 Carnicom, "The Expected Composition," March 28, 2002. The publisher formally asked Carnicom to remove the chart from his website.

37 A.K. Johnstone, Ph.D. *Defense Tactics: Weather Shield To Chemtrails.* Hancock House, 2002.

electromagnetic waves to follow. For example, electroactivated polymer fibers can be "pre-tuned" to conduct specific frequencies, then dropped into a barium mix to make over-the-horizon HF transmissions *bend* with the barium ducting.

From the beginning of his observations into aerosol operations, Carnicom discerned the presence of salt particles and compounds.[38] In 2001, he used a Van de Graaff generator to create sparks at repeated intervals and noticed a correlation between an increase in atmospheric conductivity and the extensive aerosol operations smearing a ghostly white throughout the once deep blue skies of New Mexico.[39] Metals are perfect for ionization, and given that moisture tends to concentrate around ions, as the Wilson cloud chamber (which detects and measures radiation) indicates, it became apparent to Carnicom that the condensation of artificial or unnatural "clouds" was being purposely created.

Given that electromagnetic energy absorption of a metallic particle is a function of particle size, conductivity, wavelength, and permeability of the vacuum constant[40], he set out to analyze exactly what was in the chemtrail mix falling on northern New Mexico. For example, daylight photo-ionizes metals like barium, but magnesium needs mid-wave UV light, UV wavelengths being just short of visible light. (Long-wave UV is the lowest frequency and a part of sunlight.) Aluminum, however, is not subject to photo-ionization[41], so was it present, too?

Eventually, he realized that these particulates were being electromagnetically zapped. In 2003, Carnicom noticed that aerosol banks of particulate concentrates close to ground level (visibility 15 miles on a sunny New Mexico day[42]) correlated with *extreme variations in local magnetic field intensity and highly pulsed VLFs*.[43] High winds appeared with aerosol banks, which pointed to thermal instabilities caused by magnetic variations in the plasma. As investigative researcher Dan Eden a.k.a. Gary Vey (www.viewzone.com) put it in "Weapons of Total Destruction" (see Appendix D, "Executive Summary," regarding "precipitation of particulates" using ELF-VLF):

> The military's own Executive Summary of the HAARP program clearly states their reliance on ELF waves. Instead of transmitting these waves from ground-based transmitters, HAARP created these waves through the use of "pulse" transmissions of their HF energy beams. Or to put it another way, HAARP duplicated the ELF signals by turning their signal on and off at rates (30 to 3000 cycles per second) within the ELF range. The result was that ELF radiation could be directed to a specific area on the surface of the planet at will.[44]

38 Clifford E. Carnicom, "Eight Conditions," September 17, 2000.
39 Carnicom, "Atmospheric Conductivity," May 20, 2001.
40 Carnicom, "Absorption Study," January 9, 2001.
41 Carnicom, "Ionization-'Clouds' Relationship," February 24, 2001; "Ionization Apparent," March 1, 2001.
42 Carnicom, "Visibility Standards Changed," March 30, 2001.
43 Carnicom, "Magnetics, Aerosols & VLF," April 13, 2003.
44 www.viewzone.com/haarp33.html

Thanks to the Internet, Carnicom and other researchers were quick to realize that the chemtrail phenomenon was not limited to the United States. Australia is heavily targeted, perhaps because it is the Down Under Echelon[45] member. In a September 10, 2012 thread, AriesJedi offers the results of an Australia rainwater sample collected in May 2010[46]:

Aluminum	720X	the safety limit
Arsenic	593X	"
Barium	300X	"
Boron	4,000X	"
Iron	2,000X	"
Manganese	4,000X	"
Zinc	8,000X	"

Another rainwater sample collected in Brisbane, Australia by Mike Scott (who works in the sterilization industry) revealed high aluminum, strontium, and barium, as well as zinc and cadmium.[47] Other condensation samples have confirmed similar results.

The following accounts for some of the traits of the lesser trace metals being dropped on us—of which industry may account for a small portion but not all—along with some of their side effects:

Silver iodide (AgI) (noted earlier) — ice nucleation for cloud seeding.

Arsenic (As) combines well with many elements, its toxicity on par with barium. Cancers and decreased production of red and white blood cells.

Boron (B) is electron-deficient and therefore reacts with chemtrail metals to make born-rich boride compounds, which are good conductors at high temperatures. Affects reproduction organs and the pineal gland.

Cadmium (Cd): Over time cadmium degenerates bones, kidneys, and lungs. (See Chapter 3.)

Iron (Fe) dispersals (such as the British Columbia "experiment" mentioned in Chapter 1) may be used for U.S. Navy sonar tests.

Lithium (Li) is a soft silver-white metal used as a desicant to dry airstreams; its ion Li+ is used as a mood stabilizer. NASA and Japan Aerospace Exploration Agency (JAXA) deployed lithium on July 4, 2013 "to create a chemical trail that can be used to track upper atmospheric winds that drive the dynamo currents. The goal is to study the global electrical current

45 The global network of NSA, GCHQ, DSD, GCSB and CSE mentioned in Chapter 1.

46 www.abovetopsecret.com/forum/thread880344/pg1

47 "Aluminum, Strontium & Barium In Brisbane Rainwater," chemtrailsnorthnz.wordpress.com/2012/10/21/aluminium-strontium-barium-in-brisbane-rainwater/

called the dynamo, which sweeps through the ionosphere, a layer of charged particles that extends from about 30 to 600 miles above Earth."[48]

Manganese (Mn): poisoning causes psychiatric, neurological, and flu-like symptoms.

Thorium (Th): With metal oxides (like manganese), thorium is capable of reflecting near-infrared heat. Radioactive thorium dioxide is a refractory Welsbach material used in nuclear fuel; causes leukemia and other cancers.

Zinc (Zn) poisoning causes stomach cramps, skin irritation, vomiting, and amnesia. High levels of zinc may mean pancreas damage, disturbed protein metabolism, and arteriosclerosis. Zinc chloride is tied to respiratory symptoms and is cumulative in fish.

Traces of chromium (Cr), gallium (Ga), lead (Pb), and selenium (Se) have also been found.

Aluminum, barium, and strontium are the most prolific hazardous metals being used to dose the atmosphere via aerosol particulates.

Aluminum (Al)

Aluminum particulates account for much of the silvery white veiling our skies now. The geoengineering/SRM theory that aluminum oxide will create an artificial sunscreen and reflect solar radiation back into space is simply not true. The truth is that with barium, aluminum is one of the silent producers of "global warming."

Francis Mangels, a 35-year wildlife biologist with the U.S. Forest Service and soil conservationist for USDA's Soil Conservation Service—now a master gardener in Mt. Shasta, California—has collected rainwater revealing 2,020 times the normal levels of aluminum, and the pH of his soil reads 7.4 instead of the normal 5.6. Aluminum, a desiccant, sucks nutrients and moisture from the soil and air and produces chlorosis (insufficient chlorophyll for photosynthesis).

Shasta has no heavy industry to blame for these high aluminum readings. The tree bark of dying California trees reveals high levels of aluminum and titanium, and as far west as Hawaii, coconut tree bark is coming off in strips while the hair analysis of a three-year-old child reads high aluminum. In Edmonton, Alberta, Canada, snowfall tests in 2002 showed elevated levels of aluminum and barium: aluminum 0.148 milligrams/liter, barium 0.006 milligrams/liter.[49]

Aluminum's high electrical conductivity makes it attractive to military agendas: "Nanoenergetic aluminum has potential military, medical, and

---

48 Ken Kremer, "NASA's Daytime Dynamo Experiment Deploys Lithium to Study Global Ionospheric Communications Disruptions." *Universe Today*, June 23, 2013.

49 Norwest Labs report #336566, November 14, 2002.

industrial applications, yet very few studies have evaluated the risk associated with these materials."[50]

Can it be true that *10–20 million tons* of aluminum might be in the upper atmosphere?

> Scientist David Keith, standing aside [sic] fellow geoengineers Ken Caldeira and Alan Robock, said in the geoengineering seminar on February 20 that they have decided to switch their stratospheric aerosol model from sulfur to aluminum. Keith went on to say that 10–20 MEGATONS per year of aerosolized aluminum will be sprayed into the atmosphere to deflect sunlight to halt global warming. No journalist prior to [Michael] Murphy (and even now) has reported on the bait and switch by geoengineers.[51]

Aluminum strips the body of boron (B). Unassailed by aluminum, boron builds strong bones, defends the heart, and helps calcium metabolize phosphorus and magnesium. As a neurotoxin, aluminum blocks nerve impulses, dulls thinking and concentration, and can produce dizziness, memory loss, impaired coordination, involuntary tremors, speech disorders, loss of balance and energy.

In his April 18, 2012 article "Chemtrails, Nanoaluminum, Neurodegenerative and Neurodevelopmental Effects,"[52] physician Russell L. Blaylock of Belhaven University, Theoretical Neurosciences Research, LLC, addresses the nanoparticles of aluminum compounds in chemtrails aerosols and jet fuel such as FS6 trimethylaluminum, an asphyxiant added to jet fuel that is soluble in ethanol. FS6 is five times the density of air, which means it replaces oxygen.

Nanosized aluminum particles are highly reactive and induce intense inflammation in brain and spinal cord tissues, leading to Alzheimer's dementia[53] and Parkinson's and Lou Gehrig's diseases.

Dr. Blaylock stresses that airborne nanoaluminum slips up the olfactory neural tracts into the brain, passing through the brain-blood barrier and cell membranes and disrupting mitochondria. The tiny particles accumulate in the bone marrow, spleen, lymph nodes, and heart.[54] Babies, small children, and the old—those at the beginning and end of life—are most vulnerable. Given that most filtering systems cannot catch nanoaluminum, they are as

---

50 Laura K. Braydich-Stolle *et al.* "Nanosized Aluminum Altered Immune Function." *ACSNano*, Vol. 4 No. 7, July 1, 2010; Applied Biotechnology Branch, Human Effectiveness Directorate, Air Force Research Laboratory, Wright-Patterson AFB.

51 "Megatons of Aluminum to Rain Down from Global Experiment." *Idaho Observer*, March 2010.

52 www.thenhf.com/article.php?id=3298

53 Since 2000, Alzheimer's patients have increased from 4.5 million to 5.4 million in 2012. Alzheimer's Association, www.alz.org/alzheimers_disease_facts_and_figures.asp

54 Lei Chen, Robert A. Yokel, Bernard Hennig, Michaael Toborek, "Manufactured Aluminum Oxide Nanoparticles Decrease Expression of Tight Junction Proteins in Brain Vasculature." *PMC*, October 1, 2008.

vulnerable indoors as outdoors. Absorbed into the GI tract in high quantities, nanoaluminum inflames the lungs, which leads to asthma and pulmonary diseases. Dr. Blaylock ends his article with an appeal to pilots and political officials to assess what they are subjecting their own families to, if they aren't concerned about other people's families.

Finally, a word about aluminum fluoride. Is the political push to infuse every municipal water supply with fluoride about more than unloading fluorosilicic acid, which, according to the EPA, is a toxic pollutant? If cities and dentists weren't pushing for fluoridation, fertilizer corporations would have to spend $4,000 to $7,000 per tanker truck just to get rid of it.[55] Finally, an ominous insight by Romanian-born Kevin Mugur Galalae in his book *Why the water and milk we drink and the salt we eat make us infertile, feeble-minded and ill*:

> Aluminum binds with fluoride to form aluminum fluoride compounds that greatly increase fluoride toxicity . . .In other words, you can do far more damage to human health with aluminum fluoride than you can do with just fluoride, and you need less of it.[56]

Barium (Ba)

Like aluminum, barium is silvery white. Together, the two metals work to diffuse and strengthen electrical charges, much as iron and copper might when connected by an acid.

In late 2000, by means of a diffraction grating spectrometer (determines matter's absorption and emission of light and other radiation), Carnicom confirmed that radical atmospheric changes induced by aircraft aerosol operations included the presence of signature high-intensity spectral lines matching barium salt compounds. *Barium was being introduced into the atmosphere on a massive scale.*[57]

Because barium reacts with almost all nonmetals, it forms many water-soluble and acid-soluble compounds (especially with sulfur, carbon, and oxygen): chrome yellow (barium chromate) barium chlorate and nitrate, barium chloride, the barium sulfide that phosphoresces, etc. Barite (barium sulfate), a sulfate ore, is insoluble in water and acids and is therefore used for internal body X-rays. The compounds *barium carbonate, barium oxide* (the drying agent that absorbs carbon dioxide and water), *barium hydroxide,* and *barium hydrate* meet the following conditions, observations, and analyses:

---

55 "Cheese is chalk if fluoride is fluoride says Hamilton oncologist." Press release, Fluoride Free Hamilton, June 28, 2013.

56 *Shift Frequency*, October 10, 2012.

57 Carnicom, "Barium Affirmed by Spectroscopy," November 1, 2000; "Barium Identification Further Confirmed," November 28, 2000.

(1) A salt crystal; absorbs moisture at low levels of relative humidity, i.e. hygroscopic;
(2) Expected to be soluble;
(3) Reactive with water but not explosive;
(4) Reacts with cold water;
(5) Alkaline in nature when combined with water;
(6) Provides unique spectrometry signature in the visible light range identified with a specific element;
(7) Ionizable, as evidenced in particulate imagery;
(8) Colorless or white;
(9) Electrolytic in nature, i.e. subject to disassociation of ions in water;
(10) Microwave frequencies are subject to disruption with injection of particles into the atmosphere;
(11) Has an estimated vapor pressure of approximately .0143torr at -50°C;
(12) Historical interest and experimentation documented with use of element(s) in ionization and plasma physics;
(13) Respiratory distress associated with ingestion into the respiratory tract;
(14) Highly probable it involves a product of combustion;
(15) Favorable conditions for aerosol dispersion include increased moisture content and higher relative temperature.[58]

In *Nexus New Times*, Will Thomas wrote: "The barium-iron 'antennas' spread in exercises conducted out of Wright-Patterson Air Force Base act as electrolytes to enhance conductivity of radar and radio waves. Wright-Pat has been long engaged in HAARP's electromagnetic warfare program."[59] Besides the barium-iron connection, two compounds that are particularly interesting when it comes to radar are *barium titanate* and *barium stearate*. Whereas barium titanate facilitates radar studies

> . . .crystals of barium titanate, a material that can capture the pulses of certain electromagnetic frequencies in the way that a radio can pick up certain radio frequencies. When the crystal pulses, or resonates, it produces electric power.[60]

barium stearate, which is basically soap bonded to metal particulate aids high-tech 3D radar imaging like VTRPE (Variable Terrain Radio Parabolic Equation) under the U.S. Navy's Radio Frequency Mission Planner (RFMP) program mentioned earlier in this chapter. By using satellite imagery, VTRPE propagates radio waves in such a way as to generate a 3D version of terrain: battlefields, fog, clouds, apartment complexes, houses, workplaces, government offices, etc.—all

58 Clifford E. Carnicom, "The Barium Deduction," May 30, 2001.

59 Will Thomas, "Chemtrails: Covert Climate Control?" *Nexus New Times*, Nov-Dec 2001.

60 Jeane Manning, "Free Energy: Making the Impossible Possible: A New Physics for a New Energy Source." MerLIB, no date.

modeled on RFMP computer monitors. HAARP and other ionospheric heaters and "Tesla tech arrays" activate barium salts to create the scalar "ducting" needed over land. VTRPE can also *cloak* aircraft behind signal inversion layers, similar to how submarines cloak themselves from sonar beams in underwater inversion layers.

The constant alchemy of anhydrous barium oxide reacting with water produces barium hydroxide and liberates a lot of heat. (Our atmosphere's chemistry has been altered by a 20X increase in hydroxide ion concentration.) Because barium's specific heat value (0.19) is much lower than air (1.003) and water (4.184), at higher altitudes it *increases atmospheric temperature*—yet another factor in "global warming." The University of Alaska shot barium into space supposedly to study the Earth's magnetic lines, but highly reflective, silvery-white barium aids HAARP's over-the-horizon signal, given that barium's refractive index is nearly double that of water and better than glass at reflection.

> When barium reacts with water to form barium hydroxide, as it would in a moist environment, it liberates much heat. This could explain why, on heavy spray days in warm weather, people complain about the abnormal, almost microwave-type heat they feel.[61]

Like aluminum, barium is not a friend to biological life. Barium suppresses the immune system because it is absorbed rapidly through the gastrointestinal tract and deposited in the muscles, lungs, and bone. It also strips selenium (Se) from the body, which protects immunity. The EPA standard is two parts barium per million parts of drinking water[62] because the soluble salts of barium, an earth metal, are toxic and can cause difficulties in breathing, stomach and chest pains, increased blood pressure, heart arrhythmia, stomach irritation, brain swelling, muscle weakness, and damage to liver, kidneys, heart, and spleen. In fact, barium has *20 times* more chronic lethality than the worst organic chlorinated pesticide. Excessive sweating is a key sign of barium poisoning.

Louisiana CBS-affiliate KSLA reminded viewers of the 1977 Biological Testing Hearings about the U.S. Army contaminating 239 areas in the United States with biological agents in 1949 and 1969, and of the Rockefeller Report about the military's abuse of its own personnel over the last 60 years. When interviewed, director of the Poison Control Center Mark Brian said that short-term barium exposure means stomach and chest pains while chronic exposure causes blood pressure problems and wears down the immune system.

---

61 Amy Worthington, "Chemtrails: Aerosol and Electromagnetic Weapons in the Age of Nuclear War." Globalresearch.ca, June 1, 2004.

62 Section 313 of the Emergency Planning and Community Right-to-Know Act of 1986; Section 6607 of the Pollution Prevention Act of 1990.

Strontium (Sr)

Like barium, strontium is a soft silver-white alkaline earth metal and is highly reactive, its physical and chemical properties being similar to its neighbors, calcium and barium. Because barium and strontium fall in the same periodic table column as calcium and magnesium, they can and do substitute for macronutrients in your body, eventually moving into the skeletal system and soft tissues. While natural strontium is stable, its synthetic isotope present in radioactive fallout has a half-life of 28.9 years. Barium and strontium particles bond to fluoride and chloride in the upper atmosphere and are then inhaled on the Earth's surface.

With barium and niobate—an oxide of niobium[63]—strontium appears to be connected with *holography*. The Abstract of a 1998 Army Research Laboratory report reads:

> An innovative technique for generating a three-dimensional holographic display using strontium barium niobate (SBN) is discussed. The resultant image is a hologram that can be viewed in real time over a wide perspective or field of view (FOV). The holographic image is free from system-induced aberrations and has a uniform, high quality over the entire FOV. The enhanced image quality results from using a phase-conjugate read beam generated from a second photorefractive crystal acting as a double-pumped phase-conjugate mirror (DPPCM). Multiple three-dimensional images have been stored in the crystal via wavelength multiplexing.[64]

THE NON-METAL SULFUR (S)

Sulfur, once known as *brimstone* (burn stone), is the tenth most common element in the universe. Bright yellow in its elemental state, sulfur is a multivalent non-metal redox (reduction-oxidation) agent whose most distinctive property is its ability to catenate (bind to itself by forming chains). *Sulfuric acid* has many industrial uses, including fertilizers, insecticides, and fungicides.

Until geoengineering, volcanoes were the primary custodians of *native sulfur* eruptions of spewed dust, debris, and sulfur dioxide (SO_2) into the lower stratosphere, after which sulfuric raindrops would form and scatter solar

63 "The temperature stability of niobium-containing superalloys is important for its use in jet and rocket engines. Niobium is used in various superconducting materials. These superconducting alloys, also containing titanium and tin, are widely used in the superconducting magnets of MRI scanners." — Wikipedia

64 "3D Holographic Display Using Strontium Barium Niobate," Army Research Laboratory, February 1998. www.dtic.mil/cgi-bin/GetTRDoc?AD=ADA338490An

radiation—ameliorate "global warming," we would say today as manmade sulfates are being launched into the atmosphere.[65]

Though much sulfur is now being gleaned from contaminant byproducts of natural gas and oil, *elemental sulfur* is also still being synthesized by anaerobic bacteria in the salt domes along the Gulf of Mexico coast. In fact, the *Frasch process*[66] by which deep-lying sulfur is mined is all but identical to the *hydraulic fracking* creating such havoc from the Gulf and up along the New Madrid faultline (see Chapter 5).

A few of the reasons that sulfur injections might appeal to the military include how *hydrogen sulfate* acts as a semiconductor for metal salts; *sulfides* can determine pH and oxygen pressure of metal-bearing fluids; *sulfur isotopes* can identify chemicals; and tracers for hydrology (*enriched sulfur*).

The story, however, is that sulfur is entirely about ameliorating "global warming." David Keith, the Canadian environmental scientist bankrolled by Bill Gates (mentioned earlier), goes so far as to recommend dumping *sulfuric acid* (H_xSO_4), despite the "adverse effects" that seem to mean *more* "global warming":

> Recent analysis suggests that the effectiveness of stratospheric aerosol climate engineering through emission of non-condensable vapors such as SO_2 [sulfur dioxide] is limited because the slow conversion to H_xSO_4 tends to produce aerosol particles that are too large; SO_2 injection may be so inefficient that it is difficult to counteract the radiative forcing due to a CO_2 doubling. Here we describe an alternate method in which aerosol is formed rapidly *in the plume following injection of* H_xSO_4*, a condensable vapor, from an aircraft.* This method gives better control of particle size and can produce larger radiative forcing with lower sulfur loadings than SO_2 injection. Relative to SO_2 injection, it may reduce some of *the adverse effects of geoengineering such as radiative heating of the lower stratosphere. . .*[67] [Emphases added.]

A year after the paper by Keith *et al.* appeared in *Geographical Research Letters*, academics and industry leaders affiliated with Stanford University held an Energy Modeling Forum to discuss climate change and geoengineering. Lowell Wood, the "dark star protégé of Edward Teller . . .one of the Pentagon's

---

65 One such project funded by Microsoft billionaire Bill Gates' FICER (Fund for Innovative Climate and Energy Research) was to release sulfates from a balloon 15 miles above NASA's Columbia Scientific Balloon Facility at Fort Sumner, New Mexico in fall 2013. Did it happen? Balloons went up ("Last launches for NASA balloon campaign — 10/5/2013," stratocat.com.ar/news1013e.htm), but the article does not reveal if High Energy Replicated Optics (HERO) or HyperSpectral Imager for Climate Science (HySICS) entailed the release of sulfates.

66 The process involves superheating water and forcing it into the deposit to melt the sulfur which is then lifted to the surface by compressed air.

67 Pierce, J.R., D.K. Weisenstein, P. Heckendorn, T. Peter, and D.W. Keith, "Efficient formation of stratospheric aerosol for climate engineering by emission of condensable vapor from aircraft," *Geophysical Research Letters,* Vol. 37, Issue 18, September 2010.

top weaponeers and the agency's go-to guru for threat assessment and weapons development"[68]—in other words, the Hoover Institution's man—pushed spraying sulfur in the usual future tense as if it weren't already underway:

> Wood's proposal was not technologically complex. It's based on the idea, well-proven by atmospheric scientists, that volcano eruptions alter the climate for months by loading the skies with tiny particles that act as mini-reflectors, shading out sunlight and cooling the Earth. Why not apply the same principles to saving the Arctic? Getting the particles into the stratosphere wouldn't be a problem—you could generate them easily enough by burning sulfur, *then dumping the particles out of high-flying 747s, spraying them into the sky with long hoses* or even shooting them up there with naval artillery. They'd be invisible to the naked eye, Wood argued, and harmless to the environment.[69] [Emphasis added.]

Meteorologist Alan Robock—director of the meteorology undergraduate program and associate director of the Center for Environmental Prediction in the Department of Environmental Sciences at Rutgers University—criticizes anthropogenic sulfate injections from the vantage point of what chaos theory calls the *butterfly effect*: that sulfate clouds will weaken Asian and African summer monsoons and affect the crops of billions of people.[70]

Regarding the catastrophe that atmospheric sulfuric injections would necessarily be, Brian Ellis, a British engineer formerly employed by the U.N. to address ozone depletion, explains somewhat heatedly the bad math behind such proposals:

> Have you thought how much CO2 would be emitted just to lift megatons of sulphur compounds up to 20km or more? A 747 carries up to about 150,000 kg of fuel, producing 500,000 kg of CO2, to carry, say, 400 passengers and their baggage, a payload of about 36 tonnes, and this is for sub-tropopausal flight (typically 10,000 m). The moment you want to increase this to 20,000 m, you will increase the fuel requirements by 3 to 4. As a rough back-of-the-envelope calculation, you're looking at an *extra* 56 Mt [megatons] of CO2 emissions per Mt of @O2 carried up. Is that worth it? Have you thought about safety? It would need, say, 30,000 flights per year to lift 1 megaton. The aircraft would be carrying highly corrosive and toxic gases. Very few cargo-only airlines would accept transport of sulphur gases in quantities exceeding the unregulated 30 g. Imagine the tiniest leak as the pressure dropped

68 Jeff Goodell, "Can Geoengineering Save The World?" *Rolling Stone*, October 4, 2011.

69 Ibid.

70 Alan Robock, "20 reasons why geoengineering may be a bad idea." *The Bulletin*, Vol. 64, No. 2.

> with altitude, entering the pressurization system; the crew would be dead within minutes and you would risk tons of the stuff being released as the aircraft crashed. It would make Bhopal look like a Sunday School picnic if it landed in an urban area . . .*It may take a few years, but you can bet your bottom dollar that every gram of your megaton* each year *or more will end up on the earth's surface, mostly as sulphuric acid.* I don't think I need spell out the consequences to the oceans or land. In time, it would certainly cause the extinction of species. Very few species are tolerant to sulphur compounds.[71] [Emphasis added.]

Breathing sulfur dioxide (SO_2) causes severe shortness of breath and pulmonary edema. Children with asthma, the elderly, and mouth-breathing adults with chronic bronchitis and emphysema are most sensitive to it. Low concentrations even outdoors can irritate eyes, nose, throat, and the respiratory tract. Coughing, shortness of breath, headache, nausea, dizziness . . .

In short, the foregoing metal and nonmetal compounds act synergistically with any number of particulates and polymers, to which the immune system and reproductive organs are particularly vulnerable (not to mention the DNA). Neurological effects and behavioral changes, disturbance of blood circulation, heart damage, effects on eyes and eyesight, reproductive failure, stomach and gastrointestinal (GI) disorders, damage to liver and kidney functions, hearing defects, disturbances of hormonal metabolism, dermatological effects, suffocation and lung embolism. . .

And for what? Well, a variety of military agendas, but front and center is making of weather engineering an instrument of "full spectrum dominance," of war. The next and final section, "Profit and Force Multipliers," is all about how weather engineering leads the way into all kinds of cloak-and-dagger operations, from control over food production and making weather-derivative profits on Wall Street, to biological experiments riding chemtrails all the way into our blood.

---

71 Brian Ellis is referenced at various ozone protection sites, but I was unable to find the source of this quote, other than under "Swedish Politician Pernilla Hagberg raises 'chemtrail' issue in parliament" blog site at Metabunk.org.

PROFIT AND FORCE MULTIPLIERS

Folks . . .there was purple snow in Russia. PURPLE. We recently had a never before seen sky spiral in Norway, and I don't care how you try to rationalize it, that was pretty screwed up. There was a tetrahedron floating above the Kremlin on the same day, although the favorite thing to say is . . .HOAX!!! People in Chile saw the sky changing colors in the middle of the night as the earthquake destroyed their city . . .it was 3 A.M. out there, so in the blackness of night, they were watching the sky change colors where there is an absence of color (black). The same thing was viewed about 20–30 minutes before the earthquake in China.

— "Flood In Pakistan & American Secret 'HAARP Technology'???"
Pakistan Live News, September 6, 2010

CHAPTER FIVE

Exploiting Earth Changes

Research over an extended period of time indicates that there is likely a strong relationship between the appearance of the aerosol operations in a given locale and time and the interaction of the following primary variables: sunspot activity, relative humidity, change in relative humidity and the relative cloud cover. The inclusion of the solar activity within this current examination may be a significant avenue of research that establishes a series of ties with earlier discussions related to ionospheric, electromagnetic and defense projects, applications of HAARP and plasma physics . . .Current studies on planetary physics and celestial considerations may demonstrate further relationships to aerosol operations in the future.

— Clifford E. Carnicom, "Predicting the Operations: Sunspots and Humidity," October 29, 2003

Exponential biological, geological, and chemical Earth changes are underway, and it is now essential to learn how to discern the difference between what is natural and cyclic and what is anthropogenic. Geoengineering impacts wind patterns that alter ocean patterns that release methane from the floor of the ocean. It's the butterfly's wing all over again. *As above, so below.*

The Earth's rotation is wobbling with abrupt slow-downs, and the thermosphere is collapsing.[1] Radon pillars of light are rising from salt domes and sinkholes are bubbling like cauldrons across the United States, Guatemala, the Bermuda Triangle, Siberia, Antarctica. Volcanoes like Popocatepetl outside of Mexico City[2], Pavlof in Alaska[3], and Copahue in Chile[4] are on the move, and volcanic eruptions off Japan and the Canary Islands are creating islands.[5]

Seventy-five percent of America has been conditioned to believe that carbon-driven "global warming" is causing the weird weather extremes now broadly labeled "climate change." Uninformed about the "national security" technology transforming our atmosphere and bodies as it manipulates macro-weather

1 Charles Cooper, "Collapse in Earth's Upper Atmosphere Stumps Researchers." *CNET*, July 16, 2010.

2 "Popocatepetl Volcano Eruption Covers Mexican Towns in Ash." Agence France-Presse, May 8, 2013.

3 Rachel D'Oro, "Pavlof Volcano's Ash Prompts Flight Cancellations." AP, May 20, 2013.

4 "Chile, Argentina order evacuation around stirring southern volcano," Reuters, May 27, 2013.

5 Emma Thomas, "Snoopy Island! Volcanic eruption forms new landmass that resembles famous cartoon dog — complete with collar." *Daily Mail*, 31 December 2013.

systems (space weather), many succumb to fear in the face of all the crazy weather and signs and wonders occurring in the sky. People once sought solace from the magnificence of the night's starry heavens, but now light pollution exacerbated by the ionized particulates in our perennially plasma-ized atmosphere has cut off even the heavens: maximum visible magnitude is now 4 (400 stars visible) in contrast to the 6,000 stars of the previous magnitude 6.[6]

The Earth is undergoing macro-cycles *and* intentional anthropogenic change. For example, the 14 inches of "unseasonable snow" that fell on May Day 2013 in Minnesota, Wisconsin, and Wyoming: "The snow looked ready to melt away fast after hitting the ground even in the areas that saw the most accumulation on Wednesday."[7] *Was it ionized snow?* British summers are getting even colder and wetter, and the "snowbird" land of Arizona is suffering blizzard conditions.[8] In May 2013, winds of 200 mph ripped through Granbury, Texas, 65 miles southwest of Dallas. Rain and hail in South China triggered floods and landslides affecting 650,000 people. Snowmelt turned into floods in Inner Mongolia.

WARMING, OR COOLING?

Even before industrialization, the Earth cycled from warming to cooling and back again, with oceans rising and falling naturally every 20 years, though not uniformly. Sunspots vanished entirely between 1645 and 1715 during Europe's Little Ice Age, then from 1880 to 1940 industrialization heated countries up, after which the global temperature dropped a half-degree and the American Midwest suffered summer frosts. Then back up it went, with warm winters from 1973 to 1975. From the 1880s to *circa* 1950 in the Northern Hemisphere, the glaciers shrank, then grew, then shrank, and now appear to be growing again.[9]

Between 1949 and 1978, the U.S. Navy's China Lake Naval Weapons Research Center was studying hurricanes, fog, and drought. By 1958, DoD scientists were ionizing and de-ionizing charges over target areas, seeking ways to electronically modify the atmosphere. Add to that the thousands of atmospheric nuclear tests and projects like the Project West Ford attempt to produce an artificial atmosphere as a "telecommunications shield" with copper needles, but which only created an 8.5 earthquake in Alaska and loss of Chilean coastline.

By 1959—a year or so after Wilhelm Reich, MD, died at Lewisburg Federal Penitentiary and his notes on *orgone* were confiscated by the FBI—serious funding

---

6 Carnicom, "The Extinction of the Stars," June 5, 2003.

7 Matthew DeLuca, "May storm heads east after dumping up to 14 inches of snow on Midwest, Plains." *NBCNews*, May 2, 2013.

8 Michael Pearson and Steve Almasy, "30 million in path of winter storm." *CNN*, February 20, 2013.

9 Jonathan Amos, "Esa's Cryosat sees Arctic sea-ice volume bounce back." *BBC News*, 16 December 2013.

for weather modification ("climate forcing") had begun. From $3 million in 1959 to $20+ million in 1980, "climate forcing" became the quiet holy grail of the National Science Foundation, Departments of Commerce, Interior, Transportation, Agriculture and Forest Service, but most of all of the DOE and DoD. NOAA was running a severe storms lab in Oklahoma, a hurricane lab in Miami, atmospheric physics and wave propagation labs in Boulder, a geophysics lab at Princeton, and several "air resources" labs all over the country. During the Vietnam "conflict," Project Popeye covertly seeded silver iodide in Southeast Asia to wash out the Ho Chi Minh Trail.

Somewhere between cooling and warming,[10] radio science personality and former DoD climate consultant Lowell Ponte wrote *The Cooling*, with a foreword by Senator Claiborne Pell[11]. Ponte said he wrote *The Cooling* for "intuitive, emotional reasons" in the belief that Planet Earth's inhabitants "deserve a voice in how we play our hand against the cooling."

> Since the 1940s the northern half of our planet has been cooling rapidly. Already the effect in the United States is the same as if every city had been picked up by giant hands and set down more than 100 miles closer to the North Pole. If the cooling continues, warned the National Academy of Scientists in 1975, we could possibly witness the beginning of the next Great Ice Age . . .If it continues, *and no strong measures are taken to deal with it*, the cooling will cause world famine, world chaos, and probably world war, and this could all come by the year 2000.[12]

Perhaps a cooling cycle isn't the cause of today's famine, chaos, and war, but otherwise not much has changed since 1976: chemical and biological warfare experimentation on nonconsensual citizens is still going on; Southeast Asian wars are now Middle East wars; and "geophysical warfare" is now in place, thanks to ionospheric heaters like HAARP and daily Cloverleaf chemtrails. Ponte's chapter "Climatocracy: The Politics of Global Cooling" would have to be changed to "Climatocracy: The Politics of Global Warming," but the weather anomalies he wrung his hands over almost 40 years ago are a mirror image of our droughts, freakish storms and severe floods in the Mississippi Basin, Great Lakes, Pennsylvania, and New Jersey, frost, snow, and starvation in the tropics.

Of Ponte's three options—(1) change ourselves to go along with the coming "cooling," (2) each nation change their local weather/climate, or (3) global

---

10 It appears that there has been a 0.6° C rise in average global temperature over 140 years. 1918–1940, warming; 1940–1965, cooling "at precisely the time that human emissions were increasing at their greatest rate"; 1965–1970, static; 1970–1998, warming; 1998–2005, static. — Queensland geologist Bob Carter, James Cook University. "There IS a problem with global warming . . .it stopped in 1998," *The Telegraph*, 09 April 2006.

11 Senator Claiborne Pell (1918–2009) was NATO, leader of the U.S. branch of the Club of Rome, and chaired the Senate Foreign Relations Committee.

12 Lowell Ponte, *The Cooling: Has the next ice age already begun? Can we survive it?* Prentice-Hall, 1976.

control over climate—it appears that geoengineers and the military-industrial complex have opted for number three.

If it is true that what we are seeing is geoengineering experiments in skies loaded with chemtrails, it may also be true that HAARP and its *cirrus contrailus* artificial cloud cover are actually *retarding* a natural cooling cycle, thus forcing the military to keep pushing the carbons-"global warming" story to obscure the "national security" technology operating Nature's levers behind a curtain of disinformation.

In his March 2008 lecture "Weather and Climate Engineering," presented at the "Perturbed Clouds in the Climate System" forum at the Frankfurt Institute for Advanced Studies (FAS) in Germany, cloud physicist William R. Cotton (see Chapter 3) stressed (in the usual plausible deniability future tense):

> . . .cirrus clouds contribute to warming of the atmosphere owing to their contribution to downward transfer of LW [long-wave] radiation. In other words they are a greenhouse agent . . .It has even been proposed to seed in clear air in the upper troposphere to produce artificial cirrus which would warm the surface enough to reduce cold-season heating demands (Detwiler and Cho 1982). So the prospects for seeding cirrus to contribute to global surface cooling do not seem to be very good.

Five years later, *New Scientist* continued to back the charade that artificial cirrus clouds were not already being created:

> Feathery cirrus clouds are beautiful, but when it comes to climate change, they are the enemy. Found at high altitude and made of small ice crystals, they trap heat — so more cirrus means a warmer world. . .[13]

By warming the atmosphere and absorbing far more than they reflect, it is apparent that artificial cirrus clouds are not about solving greenhouse havoc. Changing relative humidity and cloud cover[14] can just as well be about inducing drought. As Carnicom puts it:

> Given that the air of the earth has a specific heat value, what would be the projected heat effect of introducing particulate forms of aluminum, barium, magnesium, titanium and calcium? The majority of these five elements will have the net effect of increasing the temperature of the atmosphere of this planet, a consequence of their specific heat values which you can find in any proper reference book.[15]

---

13 Michael Marshall, "Get cirrus in the fight against climate change." *New Scientist,* 26 January 2013.

14 Clifford E. Carnicom, "Predicting the Operations: Sunspots and Humidity," October 29, 2003; "The Aerosol Reports," May 2 and 3, 2001; "Drought Inducement," April 2, 2002; audio interview with Jeff Rense, June 4, 2002.

15 Carnicom, "Drought Inducement," April 2, 2002. *Specific heat* is the amount of heat required to

In 2009, Senators Kay Hutchinson (R-TX) and John "Jay" Rockefeller (D-WV) introduced Bill S601, "Weather Mitigation Research and Development Policy Authorization Act," to "develop and implement a comprehensive and coordinated national weather mitigation policy and a national cooperative Federal and State program of weather mitigation research and development." S601 died in committee.

In the fall of 2012, the populist UK tabloid *Daily Mail* ran the headline "Global Warming stopped 16 years ago, reveals Met Office report quietly released . . .and here is the chart to prove it." The Met Office is the UK's weather and climate change gatekeeper:

> The new data, compiled from more than 3,000 measuring points on land and sea, was issued quietly on the Internet, *without any media fanfare, and, until today, it has not been reported.*[16] [Emphasis added.]

The Met Office began furiously backpedaling about what it *hadn't* said[17], then three months later confirmed that the rise of temperature in the next five years would be 0.43°C over the 1971–2000 average—just slightly higher than the 0.4°C (.72°F) rise recorded in 1998. As Dr. David Whitehouse, science adviser of the Global Warming Policy Foundation, put it:

> That the global temperature standstill could continue to at least 2017 would mean a 20-year period of no statistically significant change in global temperatures. Such a period of no increase will pose fundamental problems for climate models. If the latest Met Office prediction is correct, then it will prove to be a lesson in humility.[18]

It also seems that the ozone hole over the Antarctic is now reported as being the second *smallest* it's been in two decades (6.9 million square miles)[19] while its concentration is the second highest. (The ozone layer is 20–30 km above the Earth, whose job it is to protect biological life from the Sun's UV radiation.) A NASA/NOAA press release claims that the cause of both the small size and high concentration is warmer temperatures in the Antarctic's lower stratosphere and "an international agreement regulating the production of certain chemicals."[20]

---

flow into a substance to produce a one-degree rise in temperature. A substance with a lower specific heat will rise in temperature with a given amount of heat than a substance with a higher specific heat.

16 David Rose, "Global Warming stopped 16 year ago, reveals Met Office report quietly released . . .and here is the chart to prove it." MailOnline, 16 October 2012.

17 For the Met's response to Rose's article, see former meteorologist Anthony Watts' site at wattsupwiththat.com/2012/10/15/the-met-office-responds-to-global-warming-stopped-16-years-ago/

18 Nick McDermott, "Global Warming has STALLED since 1998: Met Office admits Earth's temperature is rising slower than first thought." MailOnline, 8 January 2013.

19 The largest ozone hole on record (11.5 million square miles) was in 2000.

20 "2012 Antarctic Ozone Hole Second Smallest in 20 Years," NASA/NOAA press release, October 24, 2012.

And what of the carbon shibboleth? Atmospheric carbon dioxide (CO_2) is not a greenhouse gas but follows the constantly fluxing temperature between the oceans and the atmosphere. University of Winnipeg climatologist Dr. Tim Ball (drtimball.com) has written about how mainstream media ignored the publication of a Japanese Research Institute satellite map showing CO_2 data that directly contradicts the Intergovernmental Panel on Climate Change (IPCC), which claims that atmospheric residency time for CO_2 is at least one hundred years, when the actual time is five to six years. Is this part of why "global warming" was abandoned for the more encompassing "climate change," and why hypotheses ("climate models") are consistently disproven?[21] In fact, are "climate models" simply insider scoops on manmade weather events?

The catchall "climate change" demands that we continue to ask questions, given that we and future generations are subject to what is now taking place in our atmosphere and Earth. As for geoengineering natural weather fluxes and moving the jet stream around for military-industrial agendas, meteorologist Alan Robock (mentioned in Chapter 3) lays out why geoengineering is a bad idea[22]:

- Global and regional precipitation modification
- Ocean acidification
- Ozone depletion
- Effects on biosphere
- Enhanced acid precipitation
- Effects on cirrus clouds
- Whitening of the sky
- Less solar radiation for solar power
- Rapid warming
- Commercial control of technology
- Military use
- Violates current treaties
- Expensive
- Whose hand will be on the thermostat?
- How would the world agree on the optimum climate?
- Who has the moral right to modify global climate?

BIG OIL GOUGING AND HAARP TOMOGRAPHY

At first glance, it may seem that the weather in the atmosphere and the Earth changes beneath our feet are separate issues, but such is no longer the case,

21 Dr. Tim Ball, "Whether It Is Warming or Climate Change, It Cannot be the CO_2," November 9, 2011.

22 *Bulletin of the Atomic Scientists*, May/June 2008.

especially now that earth-penetrating tomography (EPT) is radiating the Earth. As we said above, the line between natural Earth changes and anthropogenic changes is not always clear and needs careful scrutiny. Certainly the energy industry (nuclear and oil) has dominated politics for so long that its ubiquitous presence is all but invisible, but it is deeply involved with the chemtrail alchemy, ELF groundwaves, and tomographic charged-particle heat of HAARP technologies.

Eighteen hundred miles below the Ring of Fire in the South Pacific Ocean, all the way up to 20°N latitude, two continent-sized thermochemical piles of rock are colliding. Another thermonuclear pile sits under the vast African continent.[23] If and when these piles merge, a massive plume will erupt. *Could the ionospheric heaters be quickening the possibility of such an eruption?*

When the U.S. Congress demanded tomographic capability of HAARP before agreeing to reinstate its funding (see Chapter 1), did they consider the blowback that would follow from bombarding the Earth with ELF waves? *Tomography* uses X-rays or ultrasound to penetrate a solid object and acquire a cross section. *Earth-penetrating tomography (EPT)* beams radio energy into the auroral electrojet, which then disperses pulsed ELF energy through "ducts" in the ionosphere to form a virtual antenna thousands of miles long. This virtual ELF antenna emits longitudinal waves that can be focused many kilometers into the Earth, depending on geology and subsurface water. Aircraft and geostationary satellite sensors collect the reflected waves and relay them to computers whose software traces the outlines of underground structures, tectonic plates and fault lines, mineral and oil deposits, etc. EPT is a powerful tool.

The transfer of pulsed ELF waves miles into the Earth through "ducts"—what electrical engineer Paul Schaefer (footnote, Chapter 3) describes as "radiation of large numbers of tiny, high-velocity particles like spinning tops"—quickens Earth processes.

Recent anomalies may or may not be connected to the constant ionizing going on. "Cobwebs" of thriving bacteria have been found in spent nuclear fuel cooling tanks 70 feet down at the Savannah River Site.[24] Five weeks after ExxonMobil's Pegasus pipeline ruptured and spewed thousands of barrels of tar sands oil in Mayflower, Arkansas, residents were still sick. Was it just the carcinogen benzene they were breathing[25], or was the benzene *syncretizing* with the chemicals and polymer particulates now in the air and Earth?

---

23 It took 4,221 seismograms and two hundred days of supercomputer simulations at the University of Utah's Center for High Performance Computing to map the lower mantle. "Deep Roots of Catastrophe: Partly molten Florida-sized blob forms atop Earth's core," February 7, 2013, phys.org/news/2013-02-deep-roots-catastrophe-partly-molten.html

24 "Not knowing the 'food' source for these bacteria adds to the mystery." Mike Gellatly, "Mystery bacteria 'cobwebs' found in SRS cooling tank." *Aiken Standard*, May 4, 2013.

25 Andrea Germanos, "Toxic Benzene Fills Air Weeks After Tar Sands Spill." *Common Dreams*, May 8, 2013.

Oil and gas gouging is going on everywhere, even as HAARP charges and recharges the chemtrailed atmosphere and Earth with ELFs. Every day, seven hundred thousand barrels of toxic tar sands oil are heading south from western Canada by rail for delivery to refineries along the Texas Gulf coast.[26] A close look at the Gulf of Mexico and the New Madrid Seismic Zone will help to clarify how natural and anthropogenic change are tightly interwoven with the energy industry—gas, oil, nuclear, and HAARP.

THE GULF OF MEXICO AND SUBSIDENCE

Basically, *subsidence* here refers to land sinking to a lower level. Subsidence is not entirely natural in that it can be quickened by practices of gas and oil acquisition as well as by HAARP tomography.

By 2100, China expects 20 percent of Dadonghai Beach in Sanya, Hainan to be under water; some of the coastline of Liaoning, Jiangsu, and Shandong provinces has already disappeared.[27] In Lakeport, California, the hilly homes of a subdivision barely 30 years old are sinking toward the bedrock 25 feet below. Cracks in roads well beyond the fill are appearing, and fissures below are opening. No one has subsidence insurance coverage.[28]

Norfolk, Virginia—home of Naval Station Norfolk, the largest U.S. Naval base—was built at sea level on reclaimed wetland and is accustomed to occasional flooding, but the present subsidence from increasingly extreme weather is another story. Norfolk has had more major storms in the past decade than in the previous four decades, according to assistant city manager Ron Williams, Jr., and it sounds like he knows more about *why* than he's willing to tell: "Williams does not call what's happening in Norfolk a symptom of climate change. 'The debate about causality we're not going to get into,' he said."[29] He worries that Norfolk won't be able to come up with the $1.6 billion it will need for reconstruction in the coming decades, but is pinning his hopes on federal block grants.

The first paragraph of the U.S. Geological Survey report, "Houston-Galveston, Texas: Managing Coastal Subsidence," nails hydraulic fracturing or "fracking" for gas and oil as causal to both flooding and subsidence:

> The greater Houston area, possibly more than any other metropolitan area in the United States, has been adversely affected by land subsidence. Extensive subsidence, caused mainly by ground-water pumping [fracking] but also by oil

---

26 Patti Domm, "Canadian oil rides south even without Keystone pipeline." *CNBC*, 4 November 2013.

27 Wang Qian, "China's sea level continues to rise." *China Daily*, February 27, 2013.

28 Tracie Cone, "One by one, homes in Calif. Subdivision sinking." AP, May 11, 2013.

29 Deborah Zabarenko, "Cities in Front Line of Adapting to Extreme Weather, Rising Sea Levels." *Claims Journal*, February 28, 2013.

> and gas extraction, has increased the frequency of flooding, caused extensive damage to industrial and transportation infrastructure, motivated major investments in levees, reservoirs, and surface-water distribution facilities, and caused substantial loss of wetland habitat.[30]

Geologist-geophysicist Jack Reed used to work for Texaco but now studies Gulf of Mexico-New Madrid geology. He says the Gulf is tectonically active and that a plate readjustment is underway; magnetic, refraction, seismic, and gravity data indicate volcanics, a northeast-trending earthquake zone, and rift zones. No less than 61 seismic points with a magnitude of 5+ run along the New Madrid, giving weight to his theory that a tectonic plate fragment in the Gulf is resonating straight up the New Madrid.

Did the British Petroleum (BP) explosion in the Gulf on April 20, 2010 resonate that tectonic plate fragment?

Four years after Hurricane Katrina struck the Gulf Coast and New Orleans (see Chapter 6), the British Petroleum (BP) oilrig *Deepwater Horizon* was drilling the deepest oil well in history 250 miles southeast of Houston at a vertical depth of 35,050 feet (10,683 meters). When the *Deepwater Horizon* platform exploded over the Macondo well at the southernmost tip of the New Madrid Fault Line, it loosed the largest oil spill in U.S. history (170 million gallons).

Then, the EPA approved the release of 3 million liters of the chemical oil dispersants Corexit EC9500A and Corexit EC9527A. Subsequent to the release, everything got worse.

> Surfrider Foundation released preliminary results of their study "State of the Beach" in which they found that Corexit appears to make it tougher for microbes to digest the oil. From the report: The use of Corexit is inhibiting the microbial degradation of hydrocarbons in the crude oil and has enabled concentrations of the organic pollutants known as PAH [polycyclic aromatic hydrocarbons] to stay above levels considered carcinogenic by the NIH [National Institutes of Health] and OSHA [Occupational Safety and Health Administration]. Through the use of 'newly developed' UV light equipment, researchers were able to detect PAHs in sand and on human skin. Corexit, they said, allows these toxins to absorb into the skin and cannot be wiped off. The mixture of Corexit and crude absorbs into wet skin faster than dry.[31]

Vast numbers of Gulf inhabitants and cleanup crews grew ill with what is now being called the Blue Flu: blood in their urine, heart palpitations, kidney and liver damage, migraines, multiple chemical sensitivity, neurological damage

30 pubs.usgs.gov/circ/circ1182/pdf/07Houston.pdf

31 From Wikipedia, "Corexit." References Julia Whitty, "BP's Corexit Oil Tar Sponged Up by Human Skin," *Rolling Stone*, April 17, 2012; and "The BP Cover-Up," *Rolling Stone*, Sept./Oct. 2010.

resulting in memory loss, rapid weight loss, respiratory system and nervous system damage, seizures, skin irritation, burning and lesions, and temporary paralysis[32]—symptoms of an immune system under assault not unlike the symptoms described in Chapters 3 and 8.

Given that BP is an American military contractor, it is not difficult to connect the dots to the possibility that Corexit was a synthetic biology "experiment" along the lines of the "experimental" release of iron sulfates off the British Columbia coast mentioned in Chapter 1. Michael Edward, a caretaker and rehabilitation specialist for a nonprofit animal rescue and sanctuary in southwest Florida, puts it this way in his excellent series on "The Gulf Blue Plague":

> A central part of the deal between BP and [the corporation] Synthetic Genomics was to create biological transfer processes for crude oil that would lead to improved recovery rates. Their goal was to develop new microbes with lab-created genomes that would improve the flow of gas and oil out of a reservoir. For an oil producer like BP, more oil and gas being recovered from a source translates into more profits. This process is known as Microbial Enhanced Oil Recovery (MEOR) . . .BP and their paid minions have released a synthetic biological plague in the Gulf of Mexico and it's out of control. The entire world is a victim of their greed and foolishness. By playing the role of creator, they have begun a very dangerous game with infinite repercussions for life as we know it.[33]

According to Edward, nickel, aluminum, and manganese have been discovered in Gulf rainwater, which is

> . . .very unusual for rain clouds originating from a saline ocean. The only logical explanation is that such elements were introduced to the Gulf water where they were drawn up by the rain clouds. *The only way that could have happened is from aerial and surface spraying and/or below surface injection* . . .Bacteria thrive in rich nutrient environments. Natural minerals are necessary building blocks for nutrients that bacteria thrive on. Think of it as hydro-fertilizing the Gulf to make it a better nutritional medium for hungry oil-eating bacterium [sic]. The so-called "dispersants" are not only breaking down the crude oil into smaller pieces, they're adding needed enhancement minerals so that the bacterium can multiply more rapidly and eat up the oil faster. Such bacteria are called Bioremediators.[34] [Emphasis added.]

---

32 "Corexit: Deadly Dispersant in Oil Spill Cleanup." *Government Accountability Project*, www.whistleblower.org/program-areas/public-health/corexit

33 Michael Edward, "The Gulf Blue Plague: It's not wise to fool Mother Nature," Rense.com, October 25, 2010. See worldvisionportal.org/wordpress/.

34 Ibid.

Corexit's synthetic microorganisms are now part of our rain cycle. The red tide algal bloom killing manatees and sea turtles in the Gulf[35] is no doubt about Corexit, as well.

Are chemtrails delivering nutrients to synthetic bacteria?

Methane may have been yet another agenda item behind the BP "accident," given that military contractors like Lockheed Martin have been interested for years in developing liquid methane for stealth aircraft like the Mach 6 Aurora replacement of the CIA's SR-71 "Blackbird."[36] Early on, Corexit microorganisms set to work on the methane lake (trillions of cubic tons!) sitting on Gulf tectonic plates to force the release of frozen hydrocarbons locked in place by immense pressures. [Hydrocarbons include organic compounds such as benzene (C_6H_6) and methane (CH_4).] The plate fragment might drop as much as 500 feet when the methane lake is finally released, at which point *every* continent will move as noxious gases like deadly hydrogen sulfide suffuse the atmosphere.

While the foam in Australia[37] and "snow rollers" in Vermont[38] indicate the chemical effects of methane, they may also point to local chemtrail "experiments" or syncretistic effects of all the chemicalized heating going on. The Arctic sea ice with its frozen hydrocarbons in its permafrost cap has also been affected. (An estimated 99 percent of gas hydrates are in ocean sediment and the remaining 1 percent in permafrost areas.) Methane hydrate or "methane ice"—the most common type of gas hydrate—is highly concentrated: one cubic foot of methane hydrate traps about 164 cubic feet of methane gas. (Specific temperatures, pressures, and an ample supply of natural gas are needed to form gas hydrates and keep them stable.)

While the U.S. Geological Survey assures us that "climate warming" will not kick off "a chain reaction of warming and methane (CH_4) releases" for a few thousand years,[39] the Arctic Methane Emergency Group (AMEG) is rightfully up in arms about the Arctic and calling for a moratorium on drilling in the Arctic.[40] Project Lucy, an aggressive Arctic geoengineering project,[41] sounds very like HAARP, with its two radio transmitters decomposing methane with two different radio frequencies. (Couple other radio frequencies with methane and nanocrystalline diamonds can be formed—a moneymaker, for sure, in an era when new diamond mines, like gold mines, are becoming scarce.[42])

---

35 "Manatees Dying in Droves on Both Coasts of Florida" — Deaths of pelicans, turtles, dolphins also increasing — "Scientists fear this is the beginning of a devastating ecosystem collapse." *ENENews*, March 29, 2013.

36 www.fas.org/irp/mystery/aurora.htm

37 "Sea Foam Covers Beach Town In Australia (VIDEO)." *Huffington Post*, January 31, 2013.

38 "Snow rollers fascinate Vermont residents" [includes video]. *CBS News*, January 22, 2013.

39 Carolyn Ruppel and Diane Noserale, "Gas Hydrates and Climate Warming — Why a Methane Catastrophe Is Unlikely." May/June 2012.

40 www.ameg.me/index.php/about-ameg

41 Malcolm Light and Sam Carana, "Project Lucy." *Arctic News*, June 4, 2012. arctic-news.blogspot.com/2012/06/project-lucy.html

42 "Project Lucy: Radio Transmitter to Decompose Methane." iowa-city-climate-advocates.org/wp-content/uploads/2012/06/ProjectLucyExtendedVersion2.pdf

Speaking of profits, in the last quarter of 2011 BP raked in $7.7 billion as Gulf shrimper catches plummeted 80 percent or more. The third generation of shrimp since the introduction of Corexit is deformed, discolored, and born without eyes or even eye sockets. "Fishermen, cleanup workers, and children report strange rashes, coughing, breathing difficulty, eye irritation and a host of other unexplained health problems that have persisted in the years since the disaster."[43]

Meanwhile, as dead dolphins wash up on the shores of Louisiana and Mississippi, HAARP ELF heat and stress pulse the Earth. . .

LOUISIANA SALT DOMES AND SINK HOLES

Louisiana is subsiding. Projections are that by the end of the 21st century, the southeast corner of the state will be under 4.3 feet of Gulf water. Coastal restoration along the Gulf after Hurricane Katrina (2005) and the *Deepwater Horizon* oil spill (2010) may not be the wisest investment, after all.[44]

Both sinkholes and earthquakes are rising and sinking all over the country. It was after Hurricane Katrina in 2005 and *Deepwater Horizon* in 2010 that a 14-acre sinkhole opened up like the pit of hell southeast of Baton Rouge and sucked down the trees and homes of Belle Rose and Bayou Corne residents. In the months before the sinkhole opened up, bubbles and the smell of gas gave fair warning to Bayou Corne. The cause? The one-by-three-mile Napoleonville salt dome being "farmed" by Texas Brine Company LLC by injecting millions of gallons of water had collapsed. (A salt dome is a large, naturally occurring underground salt deposit.) Texas Brine and Big Oil were the culprits, given that salt domes and oil are often proximate, which makes it easy to imagine how drilling, detonating, and "fracking" might trigger a sinkhole or earthquake or both.

> Natural gas filtered into the aquifer, and crude oil floated to the top of the sinkhole, about a third of a mile from the nearest homes anchored on each side of Highway 70. Louisiana officials feared explosion hazards and "potentially toxic constituents of crude oil and other hydrocarbons," though the state said continuous monitoring has detected "no hazardous concentrations." Yet earlier this month, sampling by Texas Brine found two homes with "concentrations of natural gas below the structure foundations that were above normal background levels," Assumption Parish officials reported.[45]

---

43 "BP Hauls in $7.7 Billion in Profits, Gulf Fishermen Haul in Shrimp with No Eyes." *National Resources Defense Council Switchboard*, February 17, 2012.

44 *Louisiana Weekly*, February 25, 2013.

45 Ronnie Greene, "Louisiana sinkhole shatters calm, prompts buyouts on the bayou." The Center for Public Integrity, April 15, 2013.

Once residents were evacuated, seismic monitoring detected fluid and gas movement below the two square miles surrounding the giant sinkhole. Tremors, migrating methane, and expansion hint at the 50,000,000 cubic feet of methane below and made people think more than twice about the Door Point volcano buried offshore near the sinkhole.[46]

What is not widely known is that in 1991 Texas Brine injected radioactive materials—"disposal of 'calcium and magnesium brine precipitates,' later changed to 'brine sludge,' both terms categorized as Naturally Occurring Radioactive Materials (NORM)"[47]—back into the salt caverns where they had been originally produced, thus creating a percolating pressure: "The salt dome is now suspected of causing the expanding 160,000 square feet sinkhole and natural gas venting in Bayou Corne and Grand Bayou area swamps."[48]

In fact, liquid butane and radioactive waste are being stored in salt domes all over Texas and Louisiana, which may be why mainstream media have concentrated on Florida sinkholes and ignored Louisiana. If just *one* cavern of 18.8 million barrels of butane blows, it will be 100 times Hiroshima.[49] The Louann salt dome deep in the Gulf along the shoreline is more than 200 million years old and larger than the state of Texas. Add subsidence, a sinkhole, and a mega-quake, and the Mississippi River and Gulf of Mexico could become an inland sea from New Orleans to Chicago. Just 1–2 percent of marine methane emissions can trigger a chain reaction in the *atmosphere* that can disrupt the polar jet stream—that is, beyond what HAARP is already doing to it.

Such is the power of Big Oil in America that the New Madrid continues to be "flooded, fracked, drilled, HAARPed, and tornadoed."[50] But some states are making their voices heard:

> Seven Northeastern and mid-Atlantic states announced plans Tuesday to sue the Environmental Protection Agency, saying it is violating the Clean Air Act by failing to address methane emissions from oil and gas drilling, which has boomed in nearby states such as Pennsylvania and West Virginia.
>
> New York Attorney General Eric T. Schneiderman said in a news release Tuesday that the EPA is violating the Clean Air Act by failing to address the emissions. Methane is a potent greenhouse gas, and the oil and gas industry is the largest source of emissions in this country.[51]

---

46 Deborah Dupre, "La. sinkhole emergency zone quakes, methane extends 2 miles." Examiner.com, February 28, 2013.

47 "Louisiana Sinkhole: State Documentation Reveals that Radioactive Materials Were INJECTED into Five Addition Napoleonville Salt Dome Caverns (UPDATED)." *Freedomrox*, March 18, 2013.

48 Deborah Dupre, "Bayou Sinkhole: Radioactive dome issues covered up over a year." Examiner.com, August 9, 2012.

49 Zen Gardner, "The Little Known New Madrid Pipeline Bomb," August 22, 2012, www.zengardner.com/the-little-known-new-madrid-pipeline-bomb/

50 Ibid.

51 Kevin Begos, "Drilling Methane Emissions Lawsuit: New York And 6 Other States To Sue EPA." *Huffington Post*, December 11, 2012.

THE NEW MADRID SEISMIC ZONE (FAULT LINE)

The New Madrid Seismic Zone runs along the Appalachian chain of mountains that was thrust upward eons ago when Africa butted heads with North America and caused linear thrust faults and a bulge in the asthenosphere (100–200 km below the Earth surface). As geologist-geophysicist Jack Reed alluded above, a reorganization of tectonic plates could cause major orogenic (folding and faulting of the Earth's crust) movements along the New Madrid. It's possible that Reed may have been off on the timeframe when he wrote, "If you want waterfront property, you should buy land around Indianapolis. In a couple of million years this acreage could be overlooking the Strait of America that separates western and eastern America!"[52] An inland sea from New Orleans to Chicago? Perhaps the underwater cypress forests off the Alabama Gulf coast, as revealed by Hurricane Katrina, are planning a comeback.[53]

For the story behind the present Earth activity along the New Madrid Seismic Zone bordering Illinois, Indiana, Missouri, Arkansas, Kentucky, Tennessee, and Mississippi, we have to look south to the *Deepwater Horizon* oil spill and subsequent Corexit dispersals, the tectonic plate activity in the Gulf of Mexico and its Louann salt dome, and the still-growing 26-acre Louisiana sinkhole. From the Gulf, we then look north to Clintonville, Wisconsin where, according to the USGS, the "booms" heard by many on March 19–20, 2013 were made by earthquakes 1.3 miles below.[54] Since September 2010, nearly 1,000 earthquakes have occurred in Arkansas along the New Madrid. (In 2009, Arkansas had 38 quakes.) In the same period, Oklahoma went from 50 earthquakes per year to over 1,000 in 2010 alone.

It's been two hundred years since the New Madrid blew. Now, plumes of hydrocarbon are erupting along it, due to the pressure that the methane lake sitting on top of the Gulf tectonic plate is applying. From Texas-Louisiana up the Mississippi River Valley and along the borders of New Madrid states to the Great Lakes and into the St. Lawrence Seaway, booms and earthquakes are percolating, due at least in part to the activities of five major natural gas pipelines and crude oil delivery crisscrossing the New Madrid's soft alluvial soil all the way north to Detroit, Chicago, Indianapolis, and Pittsburgh. As if pipelines were not enough, 15 nuclear reactors (not unlike the Fukushima Daiichi Nuclear Power Plant) have been built along the New Madrid. A nuclear meltdown is nightmare enough, but if the methane were to blow, the expulsion would be exponential—100 times more potent than CO_2, according to anti-geoengineering activist Dane

52 Kathy Shirley, "Gulf's Evolution Makes the Shakes." *Explorer*, November 2002. See rift map at www.aapg.org/explorer/2002/11nov/rift_map.cfm

53 Ben Raines, "Ancient forest lies 10 miles off the Alabama coast." *Press-Register*, September 2, 2012.

54 Also see "Strange sounds in the sky explained by scientists," www.bookofresearch.com/mysterious-sounds-around-the-world.htm

Wigington,[55] at which point the Mississippi River would morph into a nuclear-poisoned inland sea splitting the nation.

Is the HAARP-chemtrails pump-and-dump hurrying this Earth change along?

Earthquakes along the New Madrid occur in tandem with Big Oil's "fracking"[56] and "acid jobs,"[57] the two primary oil industry techniques. Hydraulic fracturing ("fracking") pumps pressurized water and chemicals into the ground to create microquakes in the bedrock that then produce fractures from which oil flows. Acid jobs pump chemicals like hydrofluoric acid into a well to "melt" rocks and other obstructions. So acid jobs eat away at rock and debris while fracking blasts them out of the way with water and chemicals. Is it any wonder that Earth changes follow?

Ground zero of the fracking fight is two and a half hours northwest of New York City in the little township of Dimock in Susquehanna County, Pennsylvania. Hydrofracking wells are devouring the entire county, following the usual pattern of Big Oil blackmail: first, farmland is bought up, then the old timber, then populations are displaced and fracking begins. The economic boom promised by Cabot Oil & Gas of Houston, Texas has yet to materialize. (Cabot has its own private security force.) A treatment plant on the lake now pumps chlorinated water into town because the good drinking water is contaminated with methane.[58]

Small serial earthquakes are typical of fracking along the New Madrid, and the earthquakes following radon-outgassing spikes within 72 hours often vent deadly gases. Salt domes generally contain no radon, but the rock surrounding them does, and fluctuating air pressure can pump radon into the cavity and force it to the surface.[59] In the early morning hours of the day before the Waco, Texas fertilizer plant explosion, five small earthquakes between 2.8 and 4.3 shook the ground around Oklahoma City from a depth of five miles.[60] The following month, the Pegasus tar sands oil pipeline running through Tornado Alley ruptured and released benzene into Lake Conway.[61] The list goes on.

That fracking is causal to earthquakes has been known since at least the early 1960s when the U.S. Army's Rocky Mountain Arsenal discontinued its

---

55 Wigington's background is renewable energy, before which he worked for Bechtel Power. Now a land investor and preservationist, Wigington owns a 2,000-acre wildlife preserve adjacent to Lake Shasta on which he built his off-grid home. His website is GeoEngineeringWatch.org.

56 Kevin Samson, "New Study Suggests Disturbing Link Between Fracking and Large Earthquakes." *Activist Post*, April 1, 2013.

57 Jacob Chamberlain, "Think Fracking Is Bad? Wait Until You Hear about the Gas Industry's 'Acid Jobs.'" *Common Dreams*, May 28, 2013.

58 "Dimock, PA: 'Ground Zero' In The Fight Over Fracking." stateimpact.npr.org/pennsylvania/tag/dimock/

59 Thanks to Charles Barton, *The Energy Collective*.

60 Rusty Surette, "Five Earthquakes Rattle Oklahoma, Nerves Tuesday Morning." *News 9*, April 16, 2013.

61 Tornado Alley includes northern Texas, Oklahoma, Kansas, and Nebraska. "Oklahoma tornado video: Pottawatomie County 2 fatalities reported." WPTV.com, May 20, 2013; Jon Queally, "Documents Show Exxon Lied in Aftermath of Tar Sands Pipeline Rupture." *Common Dreams*, May 22, 2013.

deep injection well of 12,045 feet due to how the fluid injection of 165 million gallons of water, waste, and chemicals triggered local underground instability that ended in earthquakes. Later studies by the U.S. Army Corps of Engineers and USGS confirmed the same, and yet fracking and acid jobs continue.

Are released frozen hydrocarbons like methane and butane behind why the New Madrid is pulling toward the northeast? *TheUnitedKnowledge* has created a telling map showing how small earthquakes from gas venting align with the booms being heard. Beginning with the earthquake northwest of Dallas on January 16, 2013 and the quake in New York a week later, *TheUnitedKnowledge* filled in the booms and shakes reported in Eatonville, Indiana, Trumbull County, Ohio, and Jamestown and Gorham, New York—not counting the quakes in Spencer, Oklahoma, Marion, Illinois, and Hawksberry, Ontario. All were along the New Madrid Fault Line.

The *booms* are reminiscent of the strange infrasound groans and hums rising from the Earth around the globe, including the Russian "Woodpecker" (1976–1989) and mid-1990s "Taos Hum" in New Mexico. Like the "thunder-like sounds" heard at Tunguska in 1908, low frequencies move through the ground in scalar waves (see Chapter 7 and Appendix J). Infrasound means electromagnetic waves along the lines of military sonar "exercises," and points to the possibility of ionospheric heaters tomographically probing the Earth and seas.

In New Zealand, mass whale beachings are almost certainly due to seismic survey vessels working for international oil and gas and using single-beam echo sounder bathymetry, side-scan sonar, sub bottom profiler, and digital seismic data-gathering instruments. Then there are the squid washing up on Santa Cruz, California beaches a few miles from the U.S. Army Presidio of Monterey, and the 18 million red-winged blackbirds that dropped as one from the sky exactly at midnight New Year's Eve 2013 in Beebe, Arkansas, 70 miles from Little Rock Air Force Base.

> "I think what we are seeing is a mass extinction of the red-winged blackbird. *Almost every able bird that could travel across North America came here tonight and died,*" [University of Arkansas at Little Rock] ornithologist Mitch Brenner says. "I believe fewer than 500 of the species remain. We will have to study further, but it is likely that the numbers are below the amount needed to sustain existence.[62] [Emphasis added.]

In 2011 in the same area, *thousands* of red-winged blackbirds dropped dead and the blame was attributed to 63,000 metric tons of Phosgene "test

62 Greg Henderson, "18 Million Blackbirds Dead on New Year's Eve in Beebe, Species Likely Extinct." *Rock City Times*, January 1, 2014.

dispersants," a poisonous gas that *explodes the lungs*. It was relocated from Iraq to the Pine Bluff Arsenal.[63]

Thus we have learned to suspect a raft of given perceptions and media hype, including "global warming" and the assumption that Earth changes are natural (or supernatural). Less visible electromagnetic technologies are merging with the oil and nuclear energy industry in service to the military-industrial complex. Beyond the assault on the Earth and seas represented by gas and oil drilling and fracking, we must now include HAARP and its earth-penetrating tomography (EPT) when it comes to deciphering what lies behind the extreme weather and Earth changes we are witnessing.

63 These tests were connected to the more than 500 measurable earthquakes in central Arkansas in 2010: Sarah Hoye, "Central Arkansas growing weary of relentless tremors." *CNN*, December 28, 2010. Also see "Top US Official Murdered After Arkansas Weapons Test Causes Mass Death," TheTotalCollapse.com, January 4, 2011.

CHAPTER SIX

Climate Engineering, Food, and Weather Derivatives

The threats of weather warfare, totalitarian government and famine dovetail together. As we saw in the famine that the Soviets artificially created in the Ukraine prior to World War II, famine is an effective means of subjugating a people. By controlling food, you can control people. Weather modification can affect food production and eventually the available supply. Starving resisters out is much more effective than having to track them down and shoot it out with them. If you have not surrendered your weapons, you don't get a food ration coupon. Long-term food storage, well hidden, is the only insulation against famine and totalitarian oppression.

— Philip L. Hoag, *No Such Thing As Doomsday: How to Prepare for Earth Changes, Power Outages, Wars & Other Threats*, 1996

As ionosphere excitation quickens carbon-heavy "global warming," failing "climate models" fail to explain away extreme weather. Either it is being intentionally engineered or it is blowback from the upper atmosphere's *reactivity* to decades of VLF power line emissions and HAARP ELF waves tampering with the jet stream (air currents), auroral electrojet (ionospheric electric current circling the Poles), and the atmosphere we breathe.

Extreme heat waves and winters, low crop yields, rising acidic sea levels, uncanny hurricanes and earthquakes, tornadoes in Colorado, sink holes and subsidence,[1] methane gas releases. In 2002, hurricane winds of 200 kilometers per hour and weeks of heavy rain buffeted Germany, France, Britain, Portugal, Belgium, Austria, the Netherlands, and the Czech Republic, while parts of the Alps remained snowless from "unseasonably warm temperatures." Belgium was under a meter of water; the German reinsurance group Munich Re said damage in Germany alone would be several hundred million euros.[2] Eleven years later, headlines continue to sound the alarm: "Storms on U.S. Plains stir

1 Lyn Leahz, "Officials investigate cause behind massive earth-split in Brazil." Vineoflife.net, March 26, 2013.

2 Erik Kirschbaum, "Devastating storms ravage Europe." *The Sun-Herald*, January 5, 2003.

memories of the 'Dust Bowl',"[3] "Taiwan Flooded with Almost 5 Feet of Typhoon Rain."[4]

Back in 1976, the National Weather Modification Policy Act was placed under the aegis of the Secretary of Commerce, perhaps due to the close relationship between weather and food production. But when the 1978 Environmental Modification Convention (ENMOD) prohibited deliberate manipulation of natural processes that sought to create severe effects on climate and weather patterns (earthquakes, tornadoes, droughts, tsunamis, etc.), the Department of Commerce made sure it had no teeth.

In 1998—the year that the chemtrails-HAARP pump-and-dump began to really rev up—the U.S. Air Force came out with *Weather As A Force Multiplier: Owning the Weather*,[5] revealing to the public that the military-industrial complex looks at weather as an instrument of "asymmetric war" for "full-spectrum dominance." (More on weather warfare in Chapter 7.) Crop failure is used to drive farmers to sell their land cheap to Big Agribiz, and devastation means profits for big contractors brought in to rebuild. The dislocated are shunted here and there by NGOs as their real estate disappears into public and private holdings. The medical-pharmaceutical complex does well, too. Whole nations can be brought to heel with economic privation or the threat thereof. Small earthquakes in tectonically sensitive zones networked with caves and tunnels can be used to carve out underground facilities and save the military a lot of work (especially with the help of a few well-placed 15,000-ton daisy-cutter bombs). War as a business model backed by weather control has become a force multiplier beyond belief.

Following in the wake of extreme weather like vultures is an entire industry dedicated to a disaster-based capitalism, rising up to profit from "climate change." On May 2, 2005, *The Nation* published Yale graduate Naomi Klein's article "The Rise of Disaster Capitalism," beginning with, "Last summer, in the lull of the August media doze, the Bush Administration's doctrine of preventive war took a major leap forward"—in reference to the State Department's new Office of the Coordinator for Reconstruction and Stabilization (OCRS). "Fittingly," Klein continued, "a government devoted to perpetual pre-emptive deconstruction now has a standing office of perpetual pre-emptive reconstruction." Three years later, her book *The Shock Doctrine: The Rise of Disaster Capitalism* documented the advent of a new kind of slippery capitalism willing to orchestrate destructive events in order to vampire profits from the targeted and traumatized.[6]

Disaster capitalism works something like this: The National Intelligence Council "forecasts" future events (coups, wars, revolutions, extreme weather, etc.)

---

3 Kevin Murphy, Reuters, December 30, 2012.

4 AP, August 2, 2012.

5 Published under *Air Force 2025*, a future study for the Air Force Chief of Staff, August 1996. See Appendix C.

6 Naomi Klein, *The Shock Doctrine: The Rise of Disaster Capitalism*. Picador, 2008.

with the OCRS coming behind to plan reconstruction as old states or regions flounder. "And if the reconstruction industry is stunningly inept at rebuilding, that may be because rebuilding is not its primary purpose . . .It's about reshaping everything."[7]

In "The Chemtrail Business" of his ebook *Chemtrails Exposed*, Peter Kirby describes how geoengineered weather fits with disaster capitalism like a hand in a glove:

> The spraying itself is carried out by the US military . . .The orders mostly come from Wall Street. The military man or men in charge of the operation take orders from an intelligence agency. The people giving orders from the intelligence agency's headquarters are taking orders from Wall Street. To make a buck, energy companies, insurance companies, big banks and other catastrophe reinsurance and weather derivatives market players direct the military in their chemtrail spraying activities. Ultimately, the taxpayer pays for it all through bailouts. Geoengineers [like David Keith of Carbon Engineering and Ken Caldeira of Intellectual Ventures] serve as consultants to both intelligence agencies and the military.

The manufacture of extreme weather, disasters, and false flag media events has become a business model, with fat cats like Big Insurance in the know about secret operations like aerosols: Donald Hart of Indianapolis wrote Clifford Carnicom in 2002 about the new "pollution [dispersal] exclusion" to his insurance policy—no coverage for pollution-related loss or any "communicable disease" loss or lawsuit.[8] Even corporate members of the World Gold Council may be investing in engineering small earthquakes so water in fault lines will vaporize and deposit gold closer to the surface.[9]

While geoengineers, the EPA, and DOE are busy selling aerosols to the public as a solar radiation palliative, the military's plan to "own the weather" looks to be just that: material gain for corporate fat cats and a weather weapon with which to muscle nations.

No more will Mother Nature be allowed her moods; now, weather fronts are sent in like battalions for global corporate profit and political advantage. A GMO agribiz food economy is being instituted and food speculators are jockeying into position. "Climate change" will play the patsy so agribiz and chemical giants can reap big profits as Wall Street traders rake in the weather derivatives.

---

7 Ibid.

8 Clifford E. Carnicom, "Insurance Exclusions," March 10, 2002.

9 Becky Oskin, "Earthquakes Turn Water Into Gold." *Our Amazing Planet*, March 19, 2013, according to the World Gold Council, an industry group.

CONTROL OVER FOOD AND SEED

In December 2012, Catherine Austin-Fitts—former Assistant Secretary of the U.S. Department of Housing and Urban Development (HUD) and now an investment advisor, banker, and entrepreneur (Solari Report Blog, solari.com/blog/)—revealed a secret on the populist *Coast to Coast AM* night radio show: government is working toward implementing a new digital currency based on food instead of the current petrodollar:

> If you look at what it is going to take to create a digital currency and really make it go globally, we need something other than oil to organize it and put it together, and I think part of the rush to control the food supply is to literally find that replacement.[10]

The rush to control the food supply . . .Weather being almost a synonym for food production, could the global economic crisis and concurrent weather crisis have something to do with a planned segue into a food (read: scarcity) economy?

To Wall Street speculators who view hunger as a commodity like anything else, HAARP-chemtrails weather manipulation must sound like the golden goose has arrived. States struggling to balance their budgets must be equally excited, if Maryland Democratic Governor Martin O'Malley's "rain tax" is any indication—a storm water management fee mandated by the EPA, calculated through satellite surveillance of private properties, and enforced locally.[11] Sunny poverty-stricken Spain is imposing a consumption tax on solar panels and wind collectors.[12] Even the rain, sun, and wind are to be taxed.

Imagine the political and economic possibilities of engineering drought. Kansas and Oklahoma wheat, cattle, and animal feed are at Dust Bowl levels as tornadoes devastate whole towns. Even the great Mississippi River superlane can be made into a mere shadow of her former might:

> A shipping superhighway that carries billions of dollars in grain, coal, steel and other commodities every year from the central United States to the Gulf of Mexico, the Mississippi [River] is near record-low levels due to the worst U.S. drought since 1956. . .[13]

---

10 Kenneth Schortgen Jr., "Austin-Fitts: Push for digital currency or a new dollar backed by food, not oil." Examiner.com, December 11, 2012. Also see Jerry Robinson's "The Coming Collapse of the Petrodollar System," *Follow the Money*, March 12, 2012.

11 Matthew Boyle, "Maryland Governor Taxes Rain." Breitbart, 10 April 2013.

12 Michael W. Dominowski, "Spain privatizes the sun." SiLive.com, August 4, 2013.

13 Tom Polansek, "Low water may halt Mississippi River transport next week." Reuters, December 27, 2012.

The American Geophysical Union is even talking about the Red and Dead Seas in 3500 BCE when a 200-year drought in Mesopotamia (Iraq) decimated the Sumerian language and ancient settlements, and populated areas shrank by 93 percent[14]—much like the Chacoans driven from the American Southwest 1125–1289 CE. In Ukraine, Bulgaria, Romania, Germany, Italy, France, Spain, and Britain, speculators, Russian oligarchs and Middle East sovereigns are grabbing up vast tracts of land that farmers can't support.

> According to research by the Transnational Institute, Via Campesina and others, half of all farmland in the EU is now concentrated in the 3% of large farms that are more than 100 hectares (247 acres) in size . . .None of the new research was done in Britain, which has some of the highest concentrations of land ownership anywhere in the world, with 70% of land reportedly owned by less than 1% of the population.[15]

Megacorporations are jockeying into position behind food. For example, the Russian/Swiss commodity trading company Glencore, the dark spawn of Marc Rich who in 1983 evaded $50 million in U.S. taxes by fleeing the country, only to be pardoned in 2001 by his buddy in elite crime, outgoing President Bill Clinton. Glencore's multibillions come from what once was called "diversification," now become little more than a euphemism for market monopoly:

> Although primarily a mining and energy company, [Glencore] has substantial interests in food—controlling around a quarter of the global market for barley, sunflower and rapeseed, and 10% of the world's wheat market . . .Last year [2010] Russia, the world's third largest wheat exporter, experienced a drought the like of which had never been recorded; fires damaged tens of thousands of acres of cereal.
>
> Glencore has now revealed its traders placed bets that the price of wheat would go up. On August 2 [2010] Glencore's head of Russian grain trading called on Russia's government to ban wheat exports. Three days later, that's what it did. The price of wheat went up by 15% in two days. . .[16]

Was HAARP "climate change" behind "the drought the like of which had never been recorded"? In the same article, the *Guardian*'s Raj Patel suggests, "Of course, just because a senior executive at one of the world's most powerful companies suggested a course of action that a country chose to follow doesn't mean Glencore made it happen." No, of course not, and the traders who placed

---

14 Tia Ghose, "Drought May Have Killed Sumerian Language." LiveScience.com, December 4, 2012.

15 John Vidal, "Land 'grabs' expand to Europe as big business blocks entry to farming." *The Guardian*, 17 April 2013.

16 Raj Patel, "At Glencore's pinnacle of capitalism, even hunger is a commodity." *The Guardian*, 5 May 2011.

bets that the price of wheat would go up didn't have inside knowledge, either. The subsequent uprising of Mozambique's starving people who couldn't afford the 15 percent hike in wheat probably didn't disturb the sleep of Glencore's then-CEO Ivan "Ten Billion Dollar Man" Glasenberg. Winners and losers, tooth and claw, survival of the fittest: the ethos of American disaster capitalism.

A corporate feudalism is sweeping the globe, intent upon returning us to the medieval ethos of land wealth and food production in the hands of the few. The petrodollar's end may indeed be in sight, and food may be shaping the next currency.

THE "GLOBAL VILLAGE" OF GIANT TRADE BLOCS

In November 2012 at the 21st East Asia Summit in Phnom Penh, Cambodia, ten member nations of ASEAN (Association of Southeast Asian Nations)—Brunei, Cambodia, Indonesia, Laos, Malaysia, Myanmar (Burma), the Philippines, Singapore, Thailand, and Vietnam—plus China, Japan, and South Korea as the ASEAN Plus Three, and the recent addition of India, Australia, New Zealand, Russia, and the United States, negotiated the Regional Comprehensive Economic Partnership. (BRICS—Brazil, Russia, India, China, and South Africa—formed in 2006 to represent developing or newly industrialized countries.)

Next came the Trans-Pacific Partnership (TPP) to counterbalance the rise of powerful Asia-Pacific alliances like ASEAN and BRICS. The Obama administration and its corporate "advisors" have just been given "fast-track authority" to finally bring TPP to fruition. (TPP was initiated under the previous Bush administration.)[17] The TPP has been contrived in utter secrecy:

> The level of secrecy surrounding the agreements is unparalleled—paramilitary teams scatter outside the premise of each round of discussions while helicopters loom overhead—media outlets impose a near-total blackout of reportage on the subject, and U.S. Senator Ron Wyden, the Chair of the Congressional Committee with jurisdiction over TPP, was denied access to the negotiation texts.[18]

Why all the secrecy? Perhaps because of the 26 chapters of the TPP draft, only two cover trade issues; otherwise, it is about how the global corporate empire will be run. It is about transnational corporate sovereignty over nations, from land use to intellectual property rights. Foreign corporations will be able to sue

---

17 Brian Winfield, Laura Litvan, and Michael C. Bender, "Congressional Deal Reached on Obama Trade-Talk Authority." *Bloomberg*, January 9, 2014: "House Ways and Means Committee Chairman Dave Camp of Michigan joined leaders of the Senate Finance Committee to offer legislation on so-called trade-promotion authority, which subjects trade deals to an up-or-down vote by Congress."

18 Nile Bowie, "The Trans-Pacific Partnership (TPP), An Oppressive US-Led Free Trade Agreement, A Corporate Power Tool of the 1%." *Global Research*, April 2, 2013.

national governments. Transnationals won't be held accountable but will be able to hold governments accountable for costs due to national laws and regulations, including health, safety, and environmental regulations. With control over the weather as an unbeatable global power chip, it is not difficult to imagine that disaster capitalism on a global scale was probably one of the topics discussed in closed sessions.

Though the TPP fine print has yet to be made public, the patent and intellectual property rights coveted by transnationals like Monsanto, Syngenta, Bayer, and Dow Chemical will no doubt include a "Plant Variety Protection" (PVP) chapter like the one imposed on Iraq by the U.S. Coalition Provisional Authority in June 2004, making it illegal for Iraqi farmers to re-use seeds harvested from "new varieties" (read: genetically modified) registered under law.[19] The PVP system is heading toward patenting life forms.[20]

Apparently, China is not welcome in the TPP,[21] but given that Aramco is teaming up with China to build an $8.5 billion oil refinery in the Red Sea port of Yanbu,[22] it could be that China with its one billion inhabitants is destined to dominate ASEAN and BRICS while the U.S. dominates TPP, the *New York Times'* theory to the contrary:

> "China's exclusion is strange, given its huge economic presence in the Asia-Pacific" region, Amitendu Palit, a visiting senior research fellow at the Institute of South Asian Studies at the National University of Singapore, wrote in a recent edition of *East Asia Forum*. "This has given rise to views that the United States is driving the Trans-Pacific Partnership with the strategic objective of marginalizing China."[23]

Iraq may be the poster child of what lies in store for nations assaulted not by NATO troops and bunker busters but by weather warfare creating disaster areas that cry out for "aid" and "reconstruction packages." Iraq's "economic reconstruction"—rhetoric of "democracy" and the "free market economy" to the contrary—required "giving foreign investors rights equal to Iraqis in exploiting Iraq's domestic market."[24]

Food sovereignty—the right to define one's own food and agriculture policies, to protect and regulate domestic agricultural production and trade,

---

19 "Iraq's new patent law: a declaration of war against farmers." Focus on the Global South and GRAIN, October 2004.

20 The Swiss corporation Nestlé now claims to own the fennel flower, *Nigella sativa*.

21 TPP's 12 members are Australia, Brunei, Canada, Chile, Japan, Malaysia, Mexico, New Zealand, Peru, Singapore, the United States, and Vietnam.

22 "Saudi oil refinery deal shows close ties." *Hu Yinan (China Daily)*, January 16, 2012.

23 Jane Perlez, "Asian Nations Plan Trade Bloc That, Unlike U.S.'s, Invites China." *New York Times*, November 20, 2012.

24 Report jointly issued by Focus on the Global South and GRAIN, October 2004, www.grain.org/es/article/entries/150-iraq-s-new-patent-law-a-declaration-of-war-against-farmers

to decide how food is to be produced and what should be grown locally and what imported[25]—is being phased out of the "global village," along with national sovereignty.

It may be that a period of weather "experimentation" is ending, or that a period of weather warfare is commencing. The U.S. government is shifting its emphasis from ameliorating "global warming" to preparing states for federal oversight of "extreme weather events" here to stay. On Halloween 2013, President Obama signed Executive Order 13653, "Preparing the United States for the Impacts of Climate Change,"[26] further centralizing federal control over U.S. weather (and no doubt the spin put on it in mainstream media). EO13653 establishes a task force composed of eight governors (including U.S. territory Guam), 14 mayors, and two local leaders:

> Officials of the EPA released a statement on [November 1] praising the order, saying it will be vital to their attempts to help local-level communities "adapt to a changing climate . . .These implementation Plans offer a roadmap for agency work to meet that responsibility, while carrying out President Obama's goal of preparing the country for climate-related challenges."[27]

Meanwhile, the public has been primed for some time to inwardly prepare for a future filled with food shortage and rising (GMO) food prices:

> Failing harvests in the US, Ukraine and other countries this year have eroded reserves to their lowest level since 1974. The US, which has experienced *record heatwaves and droughts in 2012*, now holds in reserve a historically low 6.5% of the maize that it expects to consume in the next year, says the UN . . .This year, for the sixth time in 11 years, the world will consume more food than it produces, largely because of *extreme weather* in the US and other major food-exporting countries . . .
>
> In a shocking new assessment of the prospects of meeting food needs, Lester Brown, president of the Earth policy research centre [Earth Policy Institute] in Washington, says that *the climate is no longer reliable* and the demands for food are growing so fast that a breakdown is inevitable, unless urgent action is taken.
>
> "Food shortages undermined earlier civilizations. We are on the same path. Each country is now fending for itself. The world is living one year to the next," he writes in a new book [*Full Planet, Empty Plates: The New Geopolitics of Food Scarcity*] . . ."The situation we are in is not temporary. These things will happen all the time. *Climate is in a state of flux and there is no normal any more.* We are

---

25 Ibid.

26 www.whitehouse.gov/the-press-office/2013/11/01/executive-order-preparing-united-states-impacts-climate-change

27 Perry Chiaramonte, "Obama uses executive order in sweeping takeover of nation's climate change policies." *FOX News*, November 1, 2013.

> beginning a new chapter. We will see food unrest in many more places. Armed aggression is no longer the principal threat to our future. The overriding threats to this century are *climate change*, population growth, spreading water shortages and rising food prices," Brown said.[28] [Emphases added.]

Of course, we have "climate change" to blame—or do we?

MONSANTO AND GMOS

I recognized my two selves: a crusading idealist and a cold, granitic believer in the law of the jungle.

— Edgar Monsanto Queeny,
Monsanto chairman (1943–63), *The Spirit of Enterprise*, 1934

Biodiversity is being shanghaied by a corporate monoculture in bed with the "structural adjustments" of the International Monetary Fund (IMF). The oil and genetic/chemical/pharmaceutical industries are thriving in these times of an endless "war on terror." It's a small step indeed from pesticides and herbicides to hormones, antibiotics, food additives, preservatives, colorings, artificial sweeteners, flavor enhancers, and heavy metals. Disease and symptoms of weakened immune systems are everywhere: endocrine disruption, birth defects, brain tumors, breast and prostate cancer, childhood leukemia, Alzheimer's, dementia, severe anemia, holes in GI tracts, and behavioral disorders.

From infancy to the grave, we are under attack, and that's not counting the new proteins yet to be encoded from ongoing environmental DNA alterations. Milk and soy infant formulae containing genetically engineered (GE)[29] proteins, canola and soy oils, and corn syrup are dehydrating babies who throw up and have severe diarrhea, the diagnosis being food protein-induced enterocolitis syndrome (FPIES). In the U.S., cows are injected with bovine growth hormone (rBGH), a GE hormone to increase milk production.[30]

After the worst drought in decades, followed by the wettest year on record, British farmers had no choice but to abandon time-tested farming and go to Monsanto for genetically modified (GMO) seed engineered for blight resistance.[31] As Don Westfall, a biotech industry consultant and vice president of Promar International, chortled, "The hope of the industry is that over time the

---

28 John Vidal, "UN warns of looming worldwide food crisis in 2013." *The Guardian*, 13 October 2012.
29 Genetically engineered (GE) and genetically modified organisms (GMO) are synonymous.
30 "Concern raised on impact of GE in infant formula." Voxy.co.nz, 16 April 2013.
31 Fiona Harvey and Rebecca Smithers, "Bad weather prompting more British farmers to favour GM use." *The Guardian*, 4 January 2013.

market is so flooded [with GMOs] that there's nothing you can do about it. You just sort of surrender."[32]

The Obama administration's advisor to the FDA commissioner was pivotal to getting rBGH approved in 1993 in the U.S., but it was still not allowed in Canada, Australia, New Zealand, Japan, Israel, and European Union countries.[33] Now, unfortunately, the European Commission has been muscled by the U.S. and is finally capitulating to the Monsantos of the world. A WikiLeaks cable from U.S. Ambassador to France Craig Stapleton to the State Department reveals that the U.S. stooped so low as to threaten Europe with an economic "hit list" if it didn't let GMOs in:

> Europe is moving backwards not forwards on this issue with France playing a leading role, along with Austria, Italy and even the [European] Commission . . .Moving to retaliation will make clear that the current path has real costs to EU interests and could help strengthen European pro-biotech voice . . .Country Team Paris recommends that we calibrate a target retaliation list that causes some pain across the EU since this is a collective responsibility, but that also focuses in part on the worst culprits. The list should be measured rather than vicious and must be sustainable over the long term, since we should not expect an early victory. . .[34]

The European capitulation took the form of the Plant Reproductive Material Law:

> It contains restrictions on vegetables and woodland trees, as well as all other plants of any species. It will be illegal to grow, reproduce, or trade any vegetable seed or tree that has not been tested and approved by the government, more specifically the "EU Plant Variety Agency." This agency will be responsible for making a list of approved plants and an annual fee must also be forwarded to the agency if growers would like to keep what they grow on the list. The new law basically puts the government in charge of all plants and seeds in Europe, and prevents home gardeners from growing their own plants from non-regulated seeds.[35]

It is difficult to overstate Monsanto's political clout.[36] When President Obama was elected in 2008, he immediately spun the revolving corporate-government roulette wheel and rewarded Monsanto high rollers with key food

---

32 Todhunter, "Genetic Engineering and the GMO Industry," 2012.

33 Wikipedia, "Bovine somatotropin."

34 Mike Adams, "Wikileaks cable reveals U.S. conspired to retaliate against European nations if they resisted GMOs." *NaturalNews*, December 24, 2010.

35 Arjun Walla, "European Commission To Ban Heirloom Seeds and Criminalize Plants & Seeds Not Registered With Government." *Collective Evolution*, June 5, 2013.

36 Monsanto is tight with intelligence-staffed "private security" firms like Total Intelligence Solutions, a Blackwater/Erik Prince offshoot. In fact, the U.S. State Department, a Blackwater client, is practically a department of the CIA. See Jeremy Scahill, "Blackwater's Black Ops." *The Nation*, October 4, 2010.

positions: Roger Beachy, former director of the Monsanto Danforth Center, became the U.S. Department of Agriculture (USDA) director of the National Institute of Food and Agriculture; and Michael Taylor, former vice president for public policy at Monsanto, became deputy commissioner of the Food and Drug Administration (FDA). Monsanto board members are in the EPA and serve on the President's Advisory Committee for Trade Policy and Negotiations.[37] On paper, the USDA and Brazil's agricultural department may own the infamous aluminum-resistant "Terminator" seed patent, but Monsanto is the real owner. Corporations and government have become mirror images of each other, and the scientists peer-reviewing Monsanto studies and patent applications are in bed with the biotech industry.

Besides dominating the $13.3 billion biotech seed industry with its herbicide Roundup and 3,981 abiotic stress-tolerant transgenic plant patents, Monsanto seems to have the inside track on chemtrail heavy metals and HAARP ionization of the atmosphere, its abiotic stress tolerance "Terminator" seed bulked up for extreme weather and aluminum-degraded soil conditions, extreme cold, heat, drought, flood, UV ozone, acid rain, salt stress, heavy metals, etc.

Monsanto and other well-endowed chemical, pharmaceutical, and genetic corporate laboratories have been pursuing *horizontal or lateral gene transfer*, a natural process for bacteria but unnatural when used to transfer genetic material from one organism to a non-offspring. One objective is to genetically engineer artificial constructs designed to cross species barriers. (See Chapter 8 for the Morgellons "hybrid.") Inserting rogue genes into genomes to make GMOs does not please Mother Nature and is therefore an unstable process, and unleashing them in the environment means even more unstable horizontal gene transfers to non-GMO organisms.

In his paper "Horizontal Gene Transfer — The Hidden Hazards of Genetic Engineering," geneticist Mae-Wan Ho[38] cautions scientists about releasing genomes into the environment, given their lack of control over horizontal gene transfers:

> Under certain ecological conditions which are still poorly understood, foreign genetic material escapes being broken down and becomes incorporated in the genome. For example, *heat shock and pollutants such as heavy metal* can favor horizontal gene transfer; and the presence of antibiotics can increase the frequency of horizontal gene transfer 10-to 10,000-fold.[39] [Emphasis added.]

---

37 Aviva Shen, "The Real Monsanto Protection Act: How the GMO Giant Corrupts Regulators and Consolidates Its Power." *ThinkProgress*, April 10, 2013.

38 Ho directs ISIS (Institute of Science in Society) that campaigns against unethical uses of biotechnology. She taught in the Department of Biological Sciences at Open University, Walton Hall, Milton Keynes, Buckinghamshire, UK.

39 ISIS Report, August 19, 2000. Also see her later report "Recent Evidence Confirms Risks of Horizontal Gene Transfer," Winter 2003. www.greens.org/s-r/30/30-14.html

The reference to "heat shock and pollutants such as heavy metal" evokes HAARP and chemtrails and the unforeseen *syncretisms* that general atmospheric heating will produce. And if the presence of antibiotics increases the frequency of horizontal gene transfer, what will PaxVax's release of a GMO cholera vaccine over Queensland, South Australia, Western Australia, and Victoria[40] do? (There has not been a case of cholera in Australia since 1977.) Abiotic stresses, transgenic antibiotic-resistant DNA . . .It may be too late to put GMOs and Monsanto's patent-mongering and aggressive genetics back into the genie bottle.

Meanwhile, laws are being tweaked and tightened so the unlawful like Monsanto can't be held accountable. Just before the April 4, 2012 Stop Trading on Congressional Knowledge Act (STOCK) was modified on April 11, 2013, the Farmers Assurance Provision (also known as the Monsanto Protection Act) was slipped into HR5973, the 2013 Agriculture, Rural Development, Food and Drug Administration, and Related Agencies Appropriations Act, to protect Monsanto and other billion-dollar giants from any and all litigation. Nor need biotech corporations worry about Environmental Impact Statements. *Food Democracy NOW!* warned that the Monsanto Protection Act would "strip judges of their constitutional mandate to protect consumer rights and the environment, while opening up the floodgates for the planting of new untested genetically engineered crops, endangering farmers, consumers and the environment."[41] The deal passed in the House in ten seconds and in the Senate in 14 seconds by voice vote only so there would be no record of who voted how.

Now and then, brave scientists try to speak up. When world-renowned expert on food safety Dr. Árpád Puztai questioned the testing of GM food during a two-and-a-half-minute interview on ITV's *World in Action*, he was fired by his employer Rowett Institute, the UK's leading food safety research lab.[42] After reading European research studies, Dr. Thierry Vrain, former head of Biotechnology at Summerland Research Station—one of Agriculture and Agri-Food Canada's national network of 19 research centers—finally woke up to the truth about GMOs:

> I refute the claims of the biotechnology companies that their engineered crops yield more, that they require less pesticide applications, that they have no impact on the environment and of course that they are safe to eat . . .There are no long-term feeding studies performed [in Canada and the U.S.] to demonstrate the claims that engineered corn and soya are safe. All we have are scientific studies out of Europe and Russia, showing that rats fed engineered food die prematurely. These studies show that proteins produced by engineered plants are different than what they should be. Inserting a gene in a genome using this technology

40 Lee Maddox, "Genetically Modified Cholera Bacteria to be Released in Australia." *Wake Up World*, 4 December 2013. Also see the Gene Technology Act 2000.

41 action.fooddemocracynow.org/sign/stop_the_monsanto_protection_act/

42 Andrew Rowell, "The Sinister Sacking of the World's Leading GM Expert — and the Trail That Leads to Tony Blair and the White House." *Daily Mail*, 7 July 2003.

> can and does result in damaged proteins. The scientific literature is full of studies showing that engineered corn and soya contain toxic or allergenic proteins.[43]

Not counting unknown "inert" additives protected as a "trade secret," the active ingredient in Roundup is *glyphosate*, secret destroyer of DNA and human placenta/umbilicus/embryo cells. Two separate studies in Sweden have linked hairy cell leukemia and non-Hodgkin lymphoma to glyphosate exposure. In the Western world, non-Hodgkin lymphoma is the most rapidly increasing cancer and in the U.S. it has risen by 73% since 1973, three years after Roundup was first introduced to the market.[44] According to internationally recognized plant pathologist Dr. Don Huber, "Glyphosate is a strong organic phosphate chelator (binds metal ions) that immobilizes positively charged minerals such as manganese, cobalt, iron, zinc, and copper . . .essential for normal physiological functions in soils, plants and animals."[45] The presence of glyphosate alone tells the tale: the nutritional benefits in natural corn are *not* supported in any way by GMO seed.

In the 2008 documentary *Food Inc.*, viewers can see how Monsanto-style agro-giants push farmers to the wall. Since 1997, Monsanto has sued 145 U.S. farmers for saving Roundup Ready soybeans and has won all 11 cases that have gone to trial. DuPont USA plays hit man for Monsanto, contracting with Agro Protection International to send ex-cops out for "farm audits" and make sure farmers are not squirreling away Roundup Ready beans for next year's planting the way they are in Argentina, where patents are not enforced.[46] DuPont and Dow Chemical contract with Monsanto for seed, pesticide, and fertilizer patents. (Monsanto controls 28 percent of the U.S. soybean market, DuPont 36 percent.) Present Roundup Ready patents owned by Monsanto will expire in the U.S. in 2014, but alternate patents are ready to take their place.

Under cover of the "war on drugs" (yet another chemical business model), Roundup is being sprayed in Colombia on subsistence crops, livestock, villages, schools, and churches in the name of eradicating the coca crops that provide the cocaine snorted and smoked by millions of Americans—and provide the CIA with the lion's share of its "bake sale" black budget.[47]

Some nations are fighting this biotech takeover of food. India is still holding on as a whole foods nation but is experiencing a devastating rate of suicide among its struggling farmers.[48] On June 5, 2012, as GMO-labeling Prop 37

---

43 "Former Pro-GMO Scientist Speaks Out on the Real Dangers of Genetically Engineered Food." *Truther*, May 7, 2013.

44 Jenny O'Connor, "Colombia's Agent Orange." *CounterPunch*, October 31, 2012.

45 "Study reveals GMO corn to be highly toxic." rt.com, April 15, 2013.

46 Jack Kashey, "DuPont Sends in Former Cops to Enforce Seed Patents: Commodities." *Bloomberg*, November 28, 2012.

47 "CIA plane crash lands with four tons of coke." *The Arnprior News*, April 14, 2011. Among many exposés, Pulitzer prize-winning journalist Gary Webb's book *Dark Alliance: The CIA, the Contras, and the Crack Cocaine Explosion* (Seven Stories Press, 1999) exposes how the CIA promotes drug addiction in America. Webb appears to have been "suicided" at 49.

48 Rubab Abid, "The myth of India's 'GM genocide': Genetically modified cotton blamed for wave of

failed in California, India's Ministry of Consumer Affairs declared that all GMO-containing packaged food must be labeled by January 1, 2013.[49] (In 2005, U.S. "permission" for India to develop nuclear power was directly linked to agreeing to the terms of the Knowledge Initiative on Agriculture drawn up by Monsanto, Cargill, and Walmart.[50])

In the United States, Connecticut and Maine have passed GMO-labeling laws.[51] On June 8 and 9, 2013, 40 tons of Syngenta's GMO sugar beets were torched in southeast Oregon on the heels of Japan's rejection of Oregon wheat because of GMO contamination. An FBI investigation considers the destruction to be "economic sabotage and a violation of federal law involving damage to commercial agricultural enterprises."[52] (In early 2012, 1,000 acres of Monsanto's GMO corn were torched in Hungary. GMO seed is illegal in Hungary.)

GEOENGINEERING AND INVESTMENTS

Millionaires don't use astrologers. Billionaires do.

— J.P. Morgan (1837–1913)

Artificial electro-based cloud formation—ELF (Extremely Low Frequency) science—will deliver a technology for stealing other nations' rain and thus increase regional tensions. Less rain/dew consequences of low sunspot numbers will be hijacked by the politicized CO_2 consensus propagandizing "science"; water-deficient nations will increase, and the importance of mutual interest water diplomacy will rise . . .Politics, beliefs, and budgets are the three main eroders of science. The CO_2 narrowly focused scientists don't understand even the most fundamental geophysical facts like the continuous shifting of both poles due to our "wobbling" globe, or the northern/southern hemisphere climate mirroring.[53]

— Gijs Graafland, Planck Foundation, 2011

farmer suicides." *National Post*, January 26, 2013: "Every 30 minutes a farmer in India commits suicide, crushed — say human-rights activists — by debt, moneylenders and the destructive policies associated with the introduction of expensive genetically modified cotton seed."

49 Anthony Gucciardi, "India Signs Mandatory GMO Labeling into Law." naturalsociety.com, January 10, 2013.

50 Colin Todhunter, "Genetic Engineering and the GMO Industry: Corporate Hijacking of Food and Agriculture." *Global Research*, December 30, 2012.

51 Michele Simon, "Connecticut Makes History as First State to Pass GE Food Labeling Law." *EcoWatch*, June 4, 2013; "Maine Governor Signs GMO Food Labeling Bill," *EcoWatch*, January 9, 2014.

52 Kimberly A.C. Wilson, "Genetically engineered sugar beets destroyed in southern Oregon." *The Oregonian*, June 20, 2013.

53 Email from Gijs Graafland, Planck Foundation. "Effects of low sunspot levels on evaporation (and by that on rain/dew/climate and health/economy)." June 24, 2011.

With the HAARP-chemtrails pump-and-dump in place, disaster capitalism is set to keep Wall Street afloat as the petro-dollar sinks and the food dollar slowly and seamlessly—for insiders, anyway—takes its place. The comprehensive infrastructure is in place, as NASA Ames Research Center physicist Minoru Freund announced in 2008, calling it an "earthquake warning system":

> Researchers say they have found a close link between *electrical disturbances* on the edge of our atmosphere and impending quakes on the ground below. Just such a signal was spotted in the days leading up to the recent devastating event in China [8.0, May 12, 2008] . . .But Minoru Freund, a physicist and director for advanced aerospace materials and devices at NASA's Ames Research Center in California, told BBC News: "I do believe that we will be able to establish a clear correlation between *certain earthquakes and certain pre-earthquake signals*, in an unbiased way." He added, "I am cautiously optimistic that we have good scientific data, *and we are designing a series of experiments to verify our data*."[54] [Emphases added.]

"Electrical disturbances at the edge of our atmosphere" and (later in the article) "a 'huge' signal in the ionosphere before the Magnitude 7.8 earthquake in Sichuan province China on 12 May" seem to be code for HAARP operations.

Working hand in glove with the "earthquake warning system" above is control over weather media. In 2011, E.L. Rothschild LLC, the private investment company of Chairman Sir Evelyn de Rothschild and CEO Lynn Forester de Rothschild, purchased a 70 percent interest in Weather Central, "the world's leading provider of interactive weather graphics and data services for television, web, and mobile."[55] Raytheon, once the owner of HAARP patents, reports the weather for the National Weather Service and NOAA through its Advanced Weather Information Processing System (AWIPS),[56] and Lockheed Martin[57] does forecasting "modeling" for the FAA. More recently, biotech giant Monsanto acquired Climate Corporation for $1 billion: "Monsanto says data science could be a $20 billion revenue opportunity beyond its core business of seeds and chemicals."[58] (Are "forecasting" and "modeling" code for "scheduling"?)

> This control of "forecasts" and "graphics" provides for the constant visual conditioning that makes the population more likely to accept the constant spraying and manipulation of our weather as "normal." When the nightly

54 Paul Rincon, "Plan for quake 'warning system'." *BBC News*, 5 June 2008.

55 "E.L. Rothschild LLC Acquires a Majority Stake in Weather Central, LP." *Business Wire*, January 31, 2011.

56 www.raytheon.com/capabilities/products/awips/ ; www.nws.noaa.gov/ops2/ops24/awips.htm

57 "The world's #1 military contractor, responsible for the U-2 and SR-71 spy planes, F-16, F/A-22 fighter jet, and Javelin missiles. They've also made millions through insider trading, falsifying accounts, and bribing officials." — *CorpWatch*

58 "Monsanto Buys Climate Corp for $930 Million," *Forbes*, October 2, 2013.

> weather "forecast animation" matches what the public sees in the sky, right down to the upper level "haze" or "overcast" from constant spraying, all seems to correlate, so nothing more is questioned. When the "forecasters" predict a hurricane will make an unprecedented 90-degree turn, and it does, nothing seems wrong. When they say there will be "heavy wet snow" at 40+ degrees, and there is, all appears OK. The global elites have, it seems, "covered all the bases."[59]

Finally in 2013—while America was distracted by terror events in Boston—the go-ahead to corporate greed was given by President Obama's signature on a partial repeal of the Stop Trading On Congressional Knowledge Act (STOCK). The searchable online finance database, the core strength of STOCK's service to the public, was removed, thus giving the nod to insider trading like weather derivatives and insider traders like Monsanto.[60]

Macro-cycles and Finance

Since the early days of Wall Street and the London Stock Exchange, cosmic macro-cycles have played a role in decisions about making money. Investors who become anxious about the latest 11-year solar cycle may check with a soothsayer like Handler & Associates, "The Option Signal Service" (www.oss.cc/oss_home.asp), or John Hampson (solarcycles.net), who offers his Internet readers these ten insights:

1. Solar cycles drive secular asset cycles.
2. Solar sunspot and geomagnetism cycles are behind economic cycles of growth and inflation.
3. Geomagnetism and lunar phasing are the key influences of market sentiment.
4. Geomagnetism trends and lunar phase oscillation create technical waveform.
5. Stock market seasonality corresponds to semi-annual geomagnetic seasonality.
6. Astro trading is based on planetary alignment which influences solar activity: the generator of sunspots and geomagnetism.
7. Historic repetition and fractals are embodied in solar and lunar cycles.
8. Demographic market models correlate with solar cycle models.
9. Lunar phasing is no more (and no less) than the influence of nocturnal illumination by the Sun.
10. A range of financial and economic disciplines manifest the same underlying solar phenomena.

59 "Rothschild's and the Geoengineering Empire," Geoengineeringwatch.org, November 13, 2012.
60 Nathan Vardi, "Did Obama and Congress Use National Security Fears to Gut the Stock Act?" *Forbes*, April 21, 2013.

Investors, like geoengineers, take a healthy interest in the cycles of our star the Sun, namely how sunspots, solar wind (plasma and charged particles), and Earth-directed CMEs (coronal mass ejections) affect investment.[61] Solar flares are blamed for people being nervous, anxious, worrisome, jittery, dizzy, shaky, irritable, lethargic, and exhausted, but also for stock market cycles that otherwise defy logic. HAARP affects our planetary energy systems much like a solar storm, and Carnicom points to a strong relationship between sunspot activity and aerosol operations in a given locale/time:

> Energy levels of the HAARP facility act on a level commensurate with solar storms. It is expected that the HAARP facility can therefore effect a global geophysical impact, including both electron density and energy state changes of the ionosphere and atmosphere.[62]

WHAT WILL HAARP MEAN FOR SOLAR CYCLES AND FINANCE?

We are now in solar cycle 24, with July 2013 being the lowest in nearly one hundred years. When the Sun moves toward a deep solar minimum, its weakening magnetic field affects earthly weather and agriculture, just as it did bankrupt Dust Bowl farmers, high food prices, and civil unrest during the last Great Depression (1929–1937). A low sunspot cycle means less air humidity, less clouds, hot summers, and cold winters. But with disaster capitalism riding high with HAARP-chemtrails geoengineering in place, even a low sunspot cycle can magically turn lead into gold. As long as "acts of God" and carbons are available as cover, the Great Game can continue one way or another, come hell, high water, or solar cycles.

Weather Derivatives

Weather derivatives are temperature-based financial instruments anyone can buy on the Chicago Mercantile Exchange (CME) or trade in over-the-counter (OTC) markets. They are usually structured as swaps, futures, and call/put options. The temperature standard of 65°F/18°C is set by the energy industry and follows atmospheric conditions reported by authorized organizations like Weather Risk Management Association (WRMA), TFS Energy, etc.

Remember Enron and the energy-deregulating scandal? Enron Weather gave birth to weather derivatives back when it basically meant offering utilities a hedge against spiking temperatures in the summer and tanking temperatures

---

61 For example, Crawford Perspectives, "a financial markets advisory service utilizing technical analysis and planetary cycles research to determine effective market-timing strategy," www.crawfordperspectives.com

62 Carnicom, "Preliminary Findings," December 17, 2003.

in the winter. Enron internationalized weather derivatives by taking it to the UK, Norway, Australia, Hong Kong, Tokyo, and Osaka. When Enron finally went bankrupt, UBS Warburg bought their trading desk *and* Enron Weather because they were so productive and stable, due no doubt to their inside track to geoengineered weather:

> . . .[Enron] employed CIA agents who could find out anything about anyone. Instead of tracking the weather on the Weather Channel, the company had a meteorologist on staff. He'd arrive at the office at 4:30 A.M., download data from a satellite, and meet with the traders at 7:00 A.M. to share his insights.[63]

Weather as market intelligence.[64] No wonder the National Weather Modification Act of 1976 was placed under the aegis of the Department of Commerce!

Enron's early days moved in tandem with HAARP coming online and the aerosol assault revving up. Coincidence? I doubt it. At activistpost.com, Kirby points out some of the history of Enron's infamous chairman and CEO Ken Lay (1942–2006). Referring to *The Smartest Guys in the Room*, Kirby reveals that Lay enlisted in the U.S. Navy in 1968 and, thanks to Enron finance chair "Pug" Winokur, was immediately transferred to the Pentagon where his "studies on the military-procurement process . . .provided the basis for his doctoral thesis on how defense spending affects the economy."[65] Lay had high-level military connections and energy-related political connections (including the Bush family) from when he worked for Exxon Mobil and when he was a Nixon aide at the Federal Power Commission and deputy undersecretary for energy at the Department of the Interior:

> A revolving door existed between Enron and the federal government. Enron executive Tom White left the company to join the Bush Jr. administration as secretary of the army . . .Robert Zoellick (now head of the World Bank) represented Enron, then the United States as Trade Representative and later as Deputy Secretary of State.[66]

From its very beginnings in 1985, Enron was in tight with top rungs of government: UK Prime Minister Margaret "Iron Lady" Thatcher (1979–1990), the infamous Reagan *troika*—President Reagan, Vice President (and former director of the CIA) George H.W. Bush, and Secretary of Defense Dick Cheney—as well as future President Bush Jr., the Pentagon, and the CIA.

Greg Palast, in his 2004 *The Best Democracy Money Can Buy: An Investigative Reporter Exposes the Truth about Globalization, Corporate Cons, and High Finance*

---

63 *Fortune* reporters Bethany McLean and Peter Elkind, *The Smartest Guys in the Room: The Amazing Rise and Scandalous Fall of Enron* (Portfolio Trade, 2004). Also, don't miss the 2005 documentary *Enron: The Smartest Guys in the Room.*

64 Peter Robison, "Hedge Funds Pluck Money From Air in $19 Billion Weather Gamble." *Bloomberg*, August 1, 2007.

65 Peter A. Kirby, "A History of Weather Derivatives." Activistpost.com, August 2, 2012.

66 Kirby, "A History of Weather Derivatives," 2012.

Fraudsters[67], underscores the Enron/Ken Lay relationship with the Bush family—a relationship as cozy as that of the Bush family with the Carlyle Group, one of the nation's largest defense contractors, specializing in private equity buyouts. Kirby was right on the money: "Chemtrail spraying is a military operation. The Carlyle Group could help geoengineering programs happen."

For a while, the good-old-boy revolving door worked well at Enron—long enough, at least, to remove all restrictions on the energy industry. (Think of how out of control gas and oil are today.) By privatizing and deregulating the energy sector, Enron and other big energy and finance players like Willis Group Holdings (terrorism risk), Koch Industries (commodities trading, ventures, investments, etc.), and PXRE Reinsurance Company were able to manipulate and arbitrage—i.e. capitalize on an imbalance between markets—remarkably well.

Heavy-hitting insider cabals of disaster capitalists are still picking the global bones left in the wake of geoengineered extreme weather. In 2011, the weather derivatives market value was $12 billion, according to the Weather Risk Management Association (WRMA). Given that 70 percent of U.S. businesses are weather-dependent (to the tune of $1 trillion per year), is it any wonder that weather engineering is such a powerful chip on the Wall Street bargaining table? The U.S., Japan, London, Amsterdam, and India may have been the initial profiteers of extreme weather, but the insider racket is slowly spreading.

The Catastrophe Reinsurance Bonanza

Beyond weather derivatives and energy futures, insurers and catastrophe reinsurers are making out like bandits, too. Compare weather derivatives' $12 billion profit in 2011 with catastrophe reinsurance's $200 billion per year. Given that the cost of Cloverleaf/Stratospheric Aerosol Program deployment is footed by taxpayers and private money like the Bill & Melinda Gates Foundation ($10 billion per year, each metric ton of aerosols costing ~$1 billion), it is obvious that catastrophe reinsurance is doing well.

Catastrophe reinsurance is the 1968 creation of the Reinsurance Association of America (RAA). Though no fan of "conflicting state and federal regulation," the RAA is interested in the "federal and state financial role for natural disaster and terrorism catastrophe risk," as well as "climate change and environmental risk."[68] Given that the RAA includes terrorism insurance, special contingency risk (kidnap and ransom), and death bonds (how many will die in a catastrophe), *blackmail* may be an unspoken specialty. Like the pirates they are, catastrophe reinsurance corporations made sure that the 1999 Financial Modernization Act passed so as to undercut the 1933 Glass-Steagall Act that separated investment from commercial banking.

Catastrophe bonds have been around since the mid-1990s and are negotiated as customized agreements found at CATEX (catastrophe risk exchange), an insider OTC (over-the-counter) transaction-matching conduit replete with secret meetings and

---

67 Plume, 2004.

68 www.reinsurance.org/Landing.aspx?id=69. For the RAA's scope, see RAA Advocacy.

homes in Bermuda, the Cayman Islands, the British Virgin Islands, Luxembourg, Ireland, etc. Once issued, "cat bonds" are divided and sold off as *securitization*: If a catastrophe results in massive claims, the issuer reserves the right to use the money behind the cat bond to pay off the claims. This alone could be why an "act of God" cover story for extreme weather must be maintained.

And so the weather derivatives game is off and running, ready to rock'n'roll from disaster to disaster, with "Terminator Seed" Monsanto a close second and astrophysicists and Wall Street astrologers alike keeping a close eye on the heavens that HAARP plays havoc with.

THE CASE OF GREECE AND CYPRUS

Thus we see how weather warfare and finance complement each other, and certainly no one has ever accused capitalism of lacking in aggression. Nations have been ruined by highly paid, well-dressed sociopathic attack dogs who, under the protection of the 13th Amendment, insist they are "corporate persons." The truth is that these collective entities with Sun Tzu's *The Art of War* in their breast pocket go for the jugular of whoever or whatever stands in their way.

In Chapter 3, I offered kudos to Europe for pressing for organized action against chemtrails. In the 1990s, Greece and Cyprus were the first to blow the whistle, claiming their right as European Community citizens to refuse to have their azure skies blotted out simply because they are NATO members. What followed was a decade of increasing economic stress. Did the chemtrails whistleblowing preempt economic sanctions taken against them?

Greece

The cover story of the 01 July 2003 issue of *Risk* magazine[69]—the same month that Aigina, Greece held the first council meeting in all of Europe to discuss chemtrail gridding and hazing—confessed that Greece's finance ministry had cut a deal with the investment bank Goldman Sachs. In 2001, Goldman Sachs had agreed to a total cross-currency swap of approximately $10 billion to keep Greece going.

The size of the upfront payment to Greece's public debt division was not clear, but what *was* clear was that "the use of derivatives in deficit and debt management by Eurozone sovereigns" is a carrot-and-stick method for making countries toe the European Union political-economic line. Add to this the threat that "the Greece-Goldman deal may be of interest to credit rating agency Standard & Poor's, which upgraded Greece's long-term debt from A to A+ in June 2003" and the objective of driving a country into a debt crisis to force privatization of natural gas, water, airports, yacht marinas, railways, and state properties becomes crystal clear.[70]

---

69 Nick Dunbar, "Revealed: Goldman Sachs' mega-deal for Greece." *Risk*, 01 July 2003.

70 "Minister: Privatizations are Greece's top priority." *USA Today*, July 7, 2012. The World Bank guarantees privatization investments so that if Greece nationalizes any of the services, corporations will file insurance claims and fleece Greece again.

The *Risk* headline followed in the wake of the mass-circulation Greek newspaper *Ethnos* lead story on Sunday, February 26, 2003, entitled, "Scientists Uneasy: Dangerous experiments in Greek Skies," with the subhead reading, "American aircraft are spraying the atmosphere with chemicals with a view to creating an artificial cloud as an 'antidote' to the Greenhouse Effect."

First, a Greek newspaper calls attention to a clandestine program being run by NATO and the U.S., and next, public exposure of Greece's quiet deal with Goldman Sachs. Then, "The European Commission initiated an excessive deficit procedure against Greece in 2004 when Greece reported an upward revision of its 2003 budget deficit figure to 3.2% of GDP."[71]

Cyprus

In 2003, neither the European Commission nor the European Central Bank felt that the prosperous Cyprus banking sector met EU standards. When Cyprus joined the EU in 2008 and began using the Euro, the country went from being fiscally sound to being bankrupt, especially when the EU began purchasing huge amounts of Greek bonds and loaning out more money than deposits could cover so as to draw Cyprus into debt slavery. In 2012, having been encouraged to support Greek bonds, Cyprus banks suffered heavy losses when the eurozone forced a restructuring of Greek sovereign debt.[72] From *EnetEnglish*, a Greek independent press:

> At least 1,600 Greek businesses — from shipping, retail to tourism — will suffer from the Cyprus bailout deal announced on Sunday after a showdown between Brussels and Nicosia [Cyprus], according to Vasilis Korkidis, head of the National Confederation of Greek Commerce (ESEE).
>
> "The tragic situation in Cyprus will certainly have immediate effects on the Greek market, since a large part of the domestic businesses maintain close ties with Cypriot companies," Korkidis said in a statement on Tuesday. He was particularly critical of the capital controls and the impending haircut on large deposits (over 100,000 euros) expected to be more than 40%.
>
> Greece's exports to Cyprus exceed 1bn euros annually and the country is Cyprus' biggest trade partner, followed by the United Kingdom and Germany.
>
> According to Korkidis, the Eurogroup's Cyprus deal establishes new, severely punitive rules for countries needing emergency aid in the future.
>
> He also slammed the Eurogroup deal (which he called the "German plan" to stress the key role played by German Chancellor Angela Merkel in the negotiations) for "crippling" Cyprus. He said the deal is "tragic" because it "sentences" Cyprus — the country's markets and economy — to a long period of recession and debt.[73]

---

71 Nelson, Rebecca M., *et al.* "Greece's Debt Crisis: Overview, Policy Responses, and Implications," *Congressional Research Service*, April 27, 2010.

72 Bruno Waterfield, "What is the 'Cyprus problem'? - Q&A." *The Telegraph*, 24 March 2013.

73 "Greek businesses to suffer from 'crippling' Cyprus bailout." *EnetEnglish*, 27 March 2013.

In Cyprus, the hated and feared savings deposit confiscation drama continues. In the Greek port city of Volos—built in 1955 over the ruins of ancient Iolcos, home of Jason who with his Argonauts boarded the *Argo* to seek the Golden Fleece—a grassroots "TEMs" barter system has been implemented[74]. Twenty-six percent of Greek children are going to school hungry, unemployment is at 27 percent, the economy in free fall. "New austerity measures" are being piled on. As Evangelina Karakaxa, 15, put it, "Those who are well fed will never understand those who are not. Our dreams are crushed. They say that when you drown, your life flashes before your eyes. My sense is that in Greece, we are drowning on dry land."[75] It is a cruel historical irony that Greece is the cradle of Western civilization.

In January 2013, economist Dimitris Kazakis—general secretary of EPAM (United People's Front)[76] and involved with ATTAC, a global justice movement—was interviewed by Konstantinos Bogdanos on the Greek television interview show *Speaking Frankly*.[77] Bogdanos referred openly to "the HAARP super-weapon" that manufactures tsunamis and earthquakes: "In the case of the tsunami in Indonesia when under Condoleezza Rice, the GNP of the United States rose by 0.8% because of the tsunami."

Kazakis agreed: "The forces of geoengineering that can destroy the planet are basically economic." He then referred to how certain hedge funds had made a billion and a half dollars from betting on Hurricane Sandy. "If I stand to make $1.7 billion from weather derivatives, why shouldn't I put a proportion of my capital into investing in the prospects of a technology that can influence the weather in one way or another?" Kazakis was clear: the capital market cannot be allowed to contract; it must continue to expand by investing in the capital itself with securities so it produces profit without the investor ever entering the real economy. Consequently, geoengineering disasters are very appealing. HAARP is an "instrument of leverage."

In economic warfare terms, what has happened to Greece and Cyprus appears to be the proverbial killing of two birds squawking too loudly about chemtrails with one stone. Finance and weather are now the weapons of choice to be wielded against nations that talk of exiting the Eurozone or don't keep official NATO secrets such as chemtrails. The lesson of Greece and Cyprus is surely not lost on world leaders watching in shock and awe as HAARP-chemtrails weather warfare advances.

---

74 Yuka Tachibana, "'People need some way out': Bartering takes hold in austerity-wracked Greece." *NBCNews*, June 15, 2012.

75 Liz Alderman, "More Children in Greece State to Go Hungry." *New York Times*, April 18, 2013.

76 epaminternational.wordpress.com/tag/dimitris-kazakis/

77 Such an unscripted show is impossible to imagine in the present United States.

CHAPTER SEVEN

Geoengineering and Environmental Warfare

Others are engaging even in an eco-type of terrorism whereby they can alter the climate, set off earthquakes, volcanoes remotely through the use of electromagnetic waves. . .

— Former Secretary of Defense William Cohen, April 1997

Technology will make available to the leaders of major nations techniques for conducting secret warfare of which only a bare minimum of the security forces need to be apprised . . .Technology of weather modification could be employed to produce prolonged periods of drought or storm.

— Zbigniew Brzezinski, *Between Two Ages: America's Role in the Techtronic Era*, 1970

ENMOD Convention prohibits the use of techniques that would have widespread, long-lasting or severe effects through deliberate manipulation of natural processes and cause such phenomena as earthquakes, tidal waves, and changes in climate and weather patterns.

— "Basic Facts About the United Nations," UN, 1994

As was claimed earlier, all things HAARP begin with Tesla.

In a 1912 interview, Nikola Tesla said it would be possible to split the planet by combining vibrations with the Earth's own resonance. "Within a few weeks, I could set the earth's crust into such a state of vibrations that it would rise and fall hundreds of feet, throwing rivers out of their beds, wrecking buildings and practically destroying civilization."[1]

Everywhere, Tesla wars are being waged against the Earth and its peoples. Extra low frequencies (ELFs)—generated by antenna arrays and coordinated with ionospheric heaters and receivers in Alaska, Russia, Norway, Puerto Rico, and others listed in Chapter 3, plus *mobile* units, like the SBX-1—are zapping chemtrail particulates for a variety of military, political, and economic purposes.

1 Allan L. Benson, "Nikola Tesla, Dreamer." *World Today*, February 1912.

The fact that our lower atmosphere is now acting less like an insulator and more like a conductor points directly to the metallic particulates being dumped by chemtrails and pumped by HAARP. Recall the introduction to Bernard Eastlund's 1987 Patent #4,686,605 deconstructed in Chapter 2:

> This invention relates to a method and apparatus for altering at least one selected region normally existing above the earth's surface and more particularly relates to a method and apparatus for altering said region by initially transmitting electromagnetic radiation from the earth's surface essentially parallel to and along naturally occurring, divergent magnetic field lines which extend from the earth's surface through the region or regions to be altered.

Now that we are going to look at various weather events that appear to be HAARP-related, let's get down to the brass tacks of how Eastlund's invention operates. Again, don't be daunted by the science or my ineptitude in explaining it; simply read more source material, observe the skies, and check out Doppler satellite feeds on the Internet.

HOW HAARP-CHEMTRAILS PUMP-AND-DUMP WORKS

Ionospheric heaters like HAARP boil the ionosphere and create a high-pressure area that pulls the stratosphere up along with the atmosphere. On both sides of the high-pressure area, a bumper zone then forces low pressure around it. In this way, the jet stream can be displaced by as much as hundreds of miles away—depending upon the length of intensity, intention, etc.—as low pressure (cold) and high pressure (hot) winds are increased to storm or hurricane levels, then steered to where the target weather is wanted. If a high-pressure zone is wanted somewhere else, HAARP, in tandem with another ionospheric heater, can move it over, say, Iran, and create a drought.

The Russian corporation Elate Intelligent Technologies and its "Weather Made to Order" director Igor Pirogoff promise that for US$200 a day, Elate hurricanes can be steered within a 200-square-mile range, and that even Hurricane Andrew could have been decreased to a "wimpy little squall."[2] Is Elate buying time from ionospheric heaters with steering capability, or is their operation small enough to simply use its own transmitters?

When Eastlund talks about intersecting beams (see Chapter 2), he is referencing *interferometry*. In fact, HAARP appears to be a scalar gravitational wave weapon, also called a *longitudinal wave interferometer*. This cutting-edge

2 Chen May Yee, "Malaysia to Battle Smog with Cyclones." *Wall Street Journal*, October 2, 1992.

quantum science has been popularized by nuclear engineer Thomas E. Bearden (cheniere.org)[3]. (See Appendix K, "Scalar Wars: The Brave New World of Scalar Electromagnetics" by Bill Morgan, 2001.) Previous incarnations of this weapon are the Tesla howitzer or Tesla Magnifying Transmitter (TMT) and the later Russian Woodpecker, both discussed in Chapter 1. According to Bearden, *scalar* refers to a quantity with magnitude or size but no motion—*mathematically* motionless but physically and inwardly in violent motion with infolded vector quantities going in all directions, as in a contained plasma moving through the fourth dimension of time in uniform motion.

Bearden, a proponent of E.T. Whittaker's superpotential theory, views how classical EM theory taught in universities is incorrect. Force fields are not primary but *potentials*, and vacuums are conglomerates of potentials. Force is mass times acceleration, which means electric force is an accelerated charged mass, not a virtual force field in vacuum. *Is physics being purposely mistaught to hide Tesla science?*

This internally trapped EM energy in local spacetime is why longitudinal (scalar) wave interferometer standing waves can work as action-at-a-distance:

> What you do is that you set up a standing wave through the earth and the molten core of the earth begins to feed that wave (we are talking Tesla now). When you have that standing wave, you have set up a triode. What you've done is that the molten core of the earth is feeding the energy and it's like your signal—that you are putting in—is gating the grid of a triode . . .Then what you do is that you change the frequency. If you change the frequency one way (start to diphase it), you dump the energy up in the atmosphere beyond the point on the other side of the earth that you focused upon. You start ionizing the air, you can change the weather flow patterns (jet streams, etc.)—you can change all of that—if you dump it gradually, real gradually—you influence the heck out of the weather. It's a great weather machine. If you dump it sharply, you don't get little ionization like that. You will get flashes and fireballs (plasma) that will come down on the surfaces of the earth . . .you can cause enormous weather changes over entire regions by playing that thing back and forth.[4]

Bearden puts it another way:

> Focus the interference zone (IZ) on the other side of the earth (beam right through the earth and ocean) to a given desired area in the atmosphere. Bias

3 Lieutenant Colonel U.S. Army (Ret.). President and CEO, CTEC, Inc.; MS in Nuclear Engineering, Georgia Institute of Technology; Graduate of Command & General Staff College and of the Guided Missile Staff Officer's Course, U.S. Army (equivalent to MS in Aerospace Engineering). A loyal military man, Bearden blames Russia and the Japanese Yakuza for what the U.S. military-industrial complex must be held responsible for. This rendering of scalar EM weapons could be supplemented by the *Megabrain Report* of 4 February 1991 with Terry Patten and Michael Hutchison at Bearden's site.

4 Lecture, U.S. Psychotronics Association Annual Conference, Dayton OH, July 25, 1981.

> your transmitters positively. You produce atmospheric heating in the air in the IZ, so that the air expands and you have produced a *low*-pressure zone. Now use a second interferometer biased negatively and place it at a distant IZ desired. In that IZ, you cool the air so it shrinks and becomes denser, and you have created a *high*-pressure area. Now place several such IZs, with the desired highs and lows, near a jet stream. The jet stream will be deviated toward a low and away from a high. By varying the transmitted energy and the IZ location (just move it gradually along), you can entrain and steer the jet streams and therefore effectively "steer" the resulting weather.[5]

Thus the interferometer creates scalar EM waves (Tesla waves, electrogravitational waves, longitudinal EM waves, waves of pure potential, electrostatic/magnetostatic waves, and zero vector waves)[6]. First, it sends out plasma orbs as target marker beacons—what people mistake for alien presences or ball lightning. Once the target is marked, the longitudinal wave is ready to bypass the 3-space world and instantaneously strike. *Scalar waves go around 3-space, not through it.*

> . . .it is possible to focus the potential for the effects of a weapon through spacetime itself, in a manner so that mass and energy do not "travel through space" from the transmitter to the target at all. Instead, *ripples and patterns in the fabric of spacetime itself* are manipulated to meet and interfere in and at the local spacetime of some distant target. There, interference of these ripple patterns creates the desired energetic effect (hence the [Russian] term energetics) directly in and through the target itself, emerging from the very spacetime (vacuum) in which the target is embedded at its distant location.[7] [Emphasis added.]

Are these "ripples and patterns in the fabric of spacetime itself" the "washboard effect" showing up in artificial cirrus clouds created by chemtrails?

Bearden does not concentrate on chemtrails, but thanks to Doppler satellite footage on the Internet, we can follow how chemtrails are laid in order to foster weather buildup and movement. Chemtrails are loaded with conductive metal particulates that further the ionizing process that the storm-steering mechanism of HAARP-like ionospheric heaters need.

Say someone wants a tornado in Joplin, Missouri, like the one on May 22, 2011. Five days before a weather front rolls in, chemtrail particulates are laid in California; two ionospheric heaters are fired up and more chemtrail particulates laid as the storm intensifies and changes direction. Make sharp bends in the jet

5 Tom Bearden, "Amazing Scalar Weather Control Reality Update." Rense.com, October 30, 1998.

6 www.cheniere.org/books/ferdelance/523.htm

7 "Energetics and Directed Energy Weapons (DEWs)," *Fer de Lance*, www.cheniere.org/books/ferdelance/intro.htm

streams and speed them up to spawn more and more little rotations; focus the IZ under the ocean to heat or cool water in a selected area.

Need something for the battlefield? Focus the IZ in a large fault zone, use the exothermic interferometry mode and deposit energy slowly to increase the stress. When the plate edges shift, bingo: an earthquake. For a really powerful quake, insert energy even more slowly so as to build overpotential, then let go. Works for volcanoes, too.

For those icebox winters that never seem to end, move an Arctic air mass south. Massive spraying over Columbus, Ohio will create a low-pressure area to draw the Canadian deep freeze south and cool areas warming too rapidly while paralyzing other areas.

Nature abhors a vacuum. Heat the plasma. Drive electrons into the lower atmosphere. Lay chemtrails to produce artificial cirrus to trap more heat. Steer the storm.

Plasma frequency is essential to the entire process. This resonant frequency of ionized gas or plasma is intimately dependent upon the electron density that determines the propagation of electromagnetic energy through the plasma[8]. Theoretical physicist Richard Feynman explains how signals from satellites to radio stations work in tandem with plasma frequency:

> This natural resonance of a plasma has some interesting effects. For example, if one tries to propagate a radio wave through the ionosphere, one finds that it can penetrate only if its frequency is higher than the plasma frequency. Otherwise the signal is reflected back. We must use high frequencies if we wish to communicate with a satellite in space. On the other hand, if we wish to communicate with a radio station beyond the horizon, we must use frequencies lower than the plasma frequency, so that the signal will be reflected back to the earth.[9]

All of this electromagnetic maneuvering impacts everything natural. The Earth and its atmosphere is a breathing organism, a whole *being* dependent upon innate rhythms, as is all of biological life here below.

> Rather than reducing the temperature of the earth and lower atmosphere, the aerosol operations commonly have the opposite effect of increasing temperature and aggravating, if not inducing, drought conditions. This is a result of a combination of factors, including the specific heats of the elements involved as well as the hygroscopic properties of the aerosols. Physical and chemical analysis of the aerosol dynamics will lead to the expected observations of increased temperature, decreased moisture and an increase in winds. The absorption and

8 Clifford E. Carnicom, "The Plasma Frequency: Radar Applications," November 5, 2001.

9 Richard Feynman, *The Feynman Lectures on Physics: Vol. II: Mainly Electromagnetism and Matter.* Addison Wesley Longman, 1970.

> displacement of this solar energy into environmental, military, biological and electromagnetic operations represents a theft of the natural and divine rights of the inhabitants of this planet. These are only preliminary effects upon the local and regional environments; longer-term and more serious impacts upon the biosphere have been, are now, and will become evident.[10]

Add to this picture supercomputers like Australia's Raijin, named after the Japanese god of thunder, lightning, and storms, which is processing millions of lines of code in order to "forecast" extreme weather.[11] Aboard the Swedish satellite *Odin*, the Canadian Space Agency's OSIRIS instrument tracks particulate aerosols in the upper atmospheres[12] as the laser beams necessary for preheating the paths of electron beams are relayed by satellite and other platforms, including nuclear submarines. ATROPATENA geophysical stations in Istanbul, Kiev, Baku, Islamabad, and Yogyakarta (Indonesia) measure the acoustic gravity waves of all the interferometric scalar tech being "tested."

It is the preheating of paths of electron beams that particularly needs conductive chemtrails. Early on, scientist Clifford Carnicom confirmed the presence of metals with a 200,000-volt Van de Graaff generator, measuring the dielectric strength or *spark* of the air whose normal dielectric strength is 3 million volts per meter (a spark length of 2.6 inches). Instead, sparks measuring 10–12 inches came up again and again, indicating that conductivity of the lower atmosphere had been altered—an increase by a factor of between three and 20 times.[13]

Two years later in 2005, by means of a calibrated meter, ammeter, ohm meter, and electrolysis in conjunction with calibrated seawater solutions, Carnicom measured the conductivity of New Mexico and Arizona snowfall samples and demonstrated conclusively

> the presence of reactive metal hydroxides (salts) in concentrations sufficient to induce visible electrolysis . . .Rainfall from such "clean" environments is not expected to support electrolysis in any significant fashion, and conductivity is expected to be on the order of 4–10μS [micro-siemens]. Current conductivity readings are in the range of approximately 15 to 25μS.[14]

An increase in conductivity of two to three times. And that's not all. The presence of submicron metal salts means that EPA air quality standards present no barrier to Cloverleaf operations. Once charged by the ohmic heating referred to in Eastlund's 1987 patent,

10 Carnicom, "The Theft of Sunlight," October 25, 2003.

11 "God of thunder: New Australian supercomputer to help predict weather." *RT*, July 31, 2013.

12 "U of S-led satellite research reveals smaller volcanoes could cool climate." University of Saskatchewan press release, July 5, 2012. Rutgers University in New Jersey, the National Center for Atmospheric Research in Colorado, and the University of Wyoming are also involved.

13 Carnicom, "Atmospheric Conductivity," July 9, 2001; "Atmospheric Conductivity II," May 7, 2003.

14 Carnicom, "Conductivity: The Air, the Water, and the Land," April 15, 2005.

> In plasma physics, ohmic heating is the energy imparted to charged particles as they respond to an electric field and make collisions with other particles. A classic definition would be the heating that results from the flow of current through a medium with electrical resistance.
>
> Metals are known to increase their resistance with the introduction of an electric current. As the metal becomes hotter, resistance increases and conductivity decreases. Salt water and plasmas are quite interesting in that the opposite effect occurs. The conductivity of salt water increases when temperature increases. The same effect occurs within a plasma; an increase in temperature will result in a decrease of the resistance, i.e. the conductivity increases. Introduction of an electric current into the plasma . . .will increase the temperature and therefore the conductivity will also increase.[15]

Increased atmospheric conductivity is inevitable. Carnicom conducted tests in his residence and at the remote Bandelier National Monument 30 miles away (to eliminate possible resonance from the 60Hz power line that feeds every residence) and ascertained a pulse width of less than 5 percent, pointing to a short spiked form of transmission in contrast to the 50 percent pulse width of a 60Hz power line signal more like a sine wave.[16]

> . . .it is well documented within the HAARP literature that ELF production that results from the pulsing of the ionosphere with high-frequency radiation is an application of importance. ELF propagation has the property of traversing extensive distances over the globe due to the extremely long wavelengths involved.[17]

And not just *over* the globe but *through* it. A final bombshell from Bearden in 1998 correspondence entitled "Scalar Electromagnetics and Weather Control":

> . . .hidden inside the matter-to-matter EM they erroneously teach us is an infolded general relativity (a matter-to-spacetime transform, followed by a spacetime-to-matter transform in serial order). This is the "magical" unified field theory everyone has been seeking. The electrodynamics we teach is a piece of tripe, and the Russians just secretly corrected it back in the early 50s . . .Today at least three other nations . . .have scalar potential interferometer weapons. The weather engineering is just one of the EARLY capabilities. *The most powerful weapons on earth are not nuclear. They are quantum potential in nature,* using a modified Whittaker EM to implement David Bohm's hidden variable theory of quantum mechanics.[18] [Emphasis added.]

---

15 Ibid.

16 Carnicom, "ELF in Bandelier National Monument," January 26, 2003.

17 Carnicom, "ELF Disruption & Countermeasures," November 26, 2002; "ELF & The Human Antenna," January 19, 2002.

18 www.tricountyi.net/~randerse/weather.htm

FORCE MULTIPLICATION

The practice of warfare, supposedly aimed at modifying the weather, we surmise to be part of certain geoengineering programs violating fundamental human rights and implemented outside of any legal framework, national or international. Although the issue has been addressed in the European Parliament (through motion B4-0551/95, demanding, among other things, transparency and public information), governments deny the reality of the spraying, impeding a necessary public debate and leaving citizens undefended. Civil Society is thus organizing itself globally to put an end to this horror that poses a threat to life and security on all the planet, polluting as it does air, soil and water.

— Josefina Fraile of Skyguards, a collaborating EU member state organization, in a letter to European Council President Demetris Christofias, September 14, 2012

Back in October 1947, the U.S. military under Project Cirrus seeded a hurricane off the Georgia coast that subsequently devastated the coastline near Savannah. The press blamed the seeding, but Nobel laureate chemist and physicist Irving Langmuir claimed it was not due to seeding but to human intervention, the purpose being to see if storms could be *steered*.[19] Experimentation with dry ice and silver iodide cloud seeding began in 1946, steering in 1947.

The final objective has always been applications of weather control beyond the local control that dry ice and silver iodide can achieve. What was needed was the capability of artificially inducing *atmospheric folding* of upper atmospheric winds from the stratosphere down into the troposphere to form a desired weather system. Accomplishing such a feat on a grand scale meant *owning* an encompassing energy delivery system and a way of focusing and steering it. In systems as dynamically unstable as global weather—in which not just air pressure and thermal systems play a part but electricity as well, particularly between the ionosphere and the troposphere—a focused *spark* in the right place at the right time would go far, but not without a method of steering. As *Angels Don't Play This HAARP* put it:

> Weather modification is possible by, for example, altering upper atmosphere wind patterns by constructing one or more plumes of atmospheric particles, which will act as a lens or focusing device. As far back as 1958, the chief White House adviser on weather modification, Capt. Howard T. Orville, said the U.S. defense department was studying "ways to manipulate the charges of the Earth

19 The Langmuir Laboratory for Atmospheric Research near Socorro, New Mexico was named after him.

> and sky and so affect the weather by using an electronic beam to ionize or de-ionize the atmosphere over a given area"...[20]

Of course, warfare was the ultimate objective behind weather control. Weather had always been a sticking point to contend with in war theatres, and now that the Revolution in Military Affairs (RMA) had reorganized military philosophy into "asymmetric" warfare, weather manipulation fit right in. Besides, force multiplication had gone just about as far as it could with nuclear tank penetrators and bunker busters[21]; now was the time for electromagnetic weapons, from "nonlethals" like Raytheon's Humvee-mounted long-range acoustic device (LRAD)[22] with a phased array antenna broadcasting a microwave beam that cooks people[23], to space lasers and particle beams and cyberwar.

Weaponizing the atmosphere fit right in.

However one chooses to look at it, the event on September 11, 2001 in New York City, one of the world's financial centers, was a game-changer. In Chapter 3, we studied the hiatus of all flights immediately after 9/11—except chemtrail flights. Just two days after 9/11, Deputy Defense Secretary Paul Wolfowitz fired a shot across the bow not just of "terrorists" but of all nations of the world:

> These people ["terrorists"] try to hide. They won't be able to hide forever ...They think their harbors are safe, but they won't be safe forever ...It's not simply a matter of capturing people and holding them accountable, but *removing the sanctuaries*, removing the support systems, *ending states* who sponsor terrorism.[24] [Emphases added.]

President George W. Bush promised that the American war on Muslims would span the globe. Eleven days after 9/11, he had a 70-minute conversation with Russian President Vladimir Putin:

> Putin gave the nod for US forces poised in Central Asia to jump into Afghanistan to be armed with tactical nuclear weapons such as small neutron bombs, which emit strong radiation, nuclear mines, shells, and other nuclear ammunition

---

20 Begich and Manning, *Angels Don't Play This HAARP*, 1995.

21 Radiation from nuclear tank penetrators and bunker busters remains in the atmosphere for 4.5 billion years. Independent scientist Leuren Moret calculated that the 2,500+ tons of depleted uranium (DU) used against Iraq in 1991 and 2003 was enough to cause 25 million new cancers, the exact population of Iraq at that time. — Leuren Moret, "The Trojan Horse of Nuclear War," paper presented at the World Depleted Uranium Weapons Conference, University of Hamburg, October 16–19, 2003. Moret's controversial vitae are at EcologyNews.com.

22 The LRAD "pain ray" was "tested" in the Middle East. The Russian "Nika" is a desktop EMP that generates *billions* of watts, according to Gennady Mesyats, director of Lebedev's Institute of Physics and vice president of the Russian Academy of Sciences. — "Russia to create electromagnetic super weapon." Pravda.com, April 9, 2008.

23 Manuel Garcia, Jr. "How to Protect Yourself From Raytheon's Pain Gun." *Counterpunch*, June 2, 2008.

24 Susanne M. Schafer, "Stage Set For Attack." AP, September 14, 2001.

> suited to commando warfare in mountainous terrain. In return, Bush assented to Russia deploying tactical nuclear weapons units around Chechnya after Moscow's ultimatum to the rebels—some of whom are backed by Osama bin Laden to surrender . . .This is an epic shift in the global balance of strength.[25]

Then began a heavy push of chemtrails-HAARP experiments. Practice runs occurred in war theatres and throughout the skies of NATO members, many of which were carried out over the U.S. "Full spectrum dominance" doctrine means military experimentation—and control—*everywhere.*

Next began a period of, well, seemingly *scheduled* earthquakes, hurricanes, and tornadoes, as observed by an anonymous observer:

> . . .has anyone noticed that almost every other year there is a new weather crisis that stays focused merely on that particular phenomenon? For instance, in 2005, the only thing that was occurring was hurricanes. You didn't hear about tornadoes, mudslides, tsunamis or volcanoes . . .it was simply hurricanes. Let's take a look at some of this stuff . . .shall we?
>
> 2005: We had Hurricanes Katrina, Dennis, Emily, Rita and Wilma. Hurricane Katrina was especially strange since it was the only hurricane on record EVER to sit for 2 days inland without any movement. This is an anomaly since it takes movement in order to keep the storm active. Not only was this the most active season in recorded history, but it also had two of the strongest measured hurricanes in recorded history, Katrina and Rita.
>
> Let's back up and look at 2004, the year of the tsunami. Remember the nonstop media coverage? It was as if nothing else existed and the only real weather was tsunami weather.
>
> 2007 brought one of the worst flood seasons in the history of mankind. No one was discussing hurricanes, tornadoes, or earthquakes. . .this year was all about floods.
>
> Now, let's look at 2008: 2008 brought the most and deadliest tornadoes in the past decade. There were no floods that year. There weren't any hurricanes happening, and if there were, no one cared because tornadoes were the main topic of interest. In almost every place in the Midwest, there were tornadoes popping up in every back alley crevice you could find. The media was all over it, and these storms were steadily increasing in power.
>
> Now let's look at the end of 2009 into 2010. EARTHQUAKES. This seems to be a fad like no other. Not only are they increasing in power, but they are also increasing in frequency. Chicago as well as Indiana, Chile (twice), Haiti, Turkey, Afghanistan, China, Okinawa, etc. . ..[26]

---

25 "Tactical nuclear weapons deployed: US and Russia strike reciprocal deal on tactical nuclear weapons deployment against Afghanistan and Chechnya." *DEBKAfile,* 6 October 2001.

26 "Flood In Pakistan & American Secret 'Haarp Technology' ???" *Pakistan Live News,* September 6,

Were these devastating weather events actually instigated by the HAARP-chemtrails force multiplier? And if they were, were they all about disaster capitalism, or were there political motivations?

EARTHQUAKES

Following Bush II's conversation with President Putin, the traumatized American people gave their president a full mandate to go to war against terrorists lurking behind every bush. At the same time that a predawn raid occurred on the southern Philippine island of Mindanao—a known haven for Muslim dissidents[27]—a M7.2 earthquake from a depth of 120 miles hit the area where U.S. forces were fighting Afghani Muslim al-Qaeda forces. Then a M7.4 earthquake in Afghanistan on March 5, and two days later a M7.6 quake on Mindanao Island, the sixth most powerful earthquake of the year. The message to nations considering harboring Muslim "terrorists" was clear to everyone but Americans, as was Wolfowitz's promise to *remove the sanctuaries* and *end states* (by weaponizing Nature).

The Russian news agency Itar-Tass was given the task of reporting that the U.S. was to blame. A source at the Geophysical Center of the Russian Academy of Sciences in Moscow was quoted as saying that beginning in the Hindu Kush and spreading to Afghanistan, Tajikistan, Uzbekistan, Pakistan, and India, the unusually strong and lengthy tremors pointed to "super-modern weapons intensively used by the U.S. aviation." From Tajik to Islamabad to New Delhi, people fled into the streets.[28]

After the M8.0 earthquake in Sichuan, China on May 12, 2008 (90,000 dead) came the M7.0 earthquake in Haiti on January 12, 2010 (500,000 dead). On ViVe TV, Venezuela President Hugo Chávez insisted that a "tectonic weapon" may have been behind the Haiti quake.[29] Exactly a month before the quake, Haiti had been the first nation to walk out of the Copenhagen Climate Change Conference after a Danish text leak revealed that Kyoto Protocol carbon tax collections would not go to developing nations but to the World Bank and IMF, and that developing nations would be forced to take out "green loans" and go further into debt.

FOX News, the right-leaning corporate mouthpiece disseminating to blue-collar America, called ViVe a "Chavez mouthpiece" and discounted its reference to the claim of the Russian Space Forces (now the Russian Aerospace Defense Forces or VKS) responsible for Russian military space operations that both

---

2010. (I removed the examples as they are no longer on the Web.)

27 Anthony Spaeth, "Rumbles in the Jungle." *Time* magazine, March 4, 2002.

28 Vladimir Suprin, "Sunday quake may have been caused by bombing." Itar-Tass, March 5, 2002.

29 Andrew Moran, "Hugo Chavez accuses U.S. of using weapon to cause Haiti quake." *Digital Journal*, January 21, 2010.

the Haiti and Christchurch earthquakes were earthquake weapon tests gone wrong. FOX then fired off the "C" word to eviscerate ViVe's forbidden allusion to HAARP:

> [ViVe TV's] Web site added that the U.S. government's HAARP program, an atmospheric research facility in Alaska (and frequent subject of conspiracy theories), was also to blame for a Jan. 9 quake in Eureka, Calif., and may have been behind the 7.8-magnitude quake in China that killed nearly 90,000 people in 2008.[30]

It is a sad state of affairs when Americans must hear their own problem-riddled news from foreign sources and not from their own "free" press. We are now on our own in America and must learn to discern truth from fiction, *whatever the source.*

Christchurch

On February 22, 2011, a M6.3 earthquake destroyed Christchurch, New Zealand (185 dead). VKS space intelligence blamed Project Wormwood at the joint Australian-U.S. Planetary Defense Base at Learmonth Solar Observatory on the North West Cape of Western Australia. Given that HAARP earthquake creation seems to follow a pump-and-dump pattern—a pumping action of, say, M4.5, which is then steered to the target some time later for a larger thrust—the Christchurch quake was on point: on February 21, the Northern Territory experienced a M4.7, then a full day later, the M6.3 hit.

If Christchurch was a target error for Tasman Glacier in the Southern Alps region 120 miles away,[31] then how are we to read the fact that a visiting American congressional delegation just happened to be in Christchurch only to fly out two and a half hours before the M6.3 hit, conveniently leaving behind FEMA deputy administrator Timothy W. Manning to coordinate the American response to the disaster they knew was coming? (General P.K. Keen, deputy commander of the U.S. Southern Command, had been similarly pre-positioned in Haiti for the 2010 devastation.[32]) Four hundred physicians just happened to be attending a urology conference at the Crown Plaza Hotel in Christchurch; they said they "felt like a wrecking ball was being smashed into the side of the building" as water poured up from the shaking ground.[33] And the 116 soldiers from the Singapore Army who just happened to be in Christchurch for a training exercise and were therefore available to cordon off the city after the quake? Coincidence?

---

30 "Hugo Chavez Mouthpiece Says U.S. Hit Haiti With 'Earthquake Weapon'." *FOX News*, January 21, 2010. Chavez died of a massive (EM?) heart attack while riddled with cancer on March 5, 2013.

31 "Christchurch quake: Ice chunks fall from Tasman Glacier." *BBC News*, 23 February 2011.

32 "US Earthquake Weapon Test Fails Again, Destroys New Zealand City." *EU Times*, February 23, 2011.

33 Belinda Merhab, "400 Aussie doctors caught in Christchurch." News.com.au, February 22, 2011.

From the same *EU Times* article:

> . . .just like Haiti was the catastrophic victim of a US "earthquake weapon" due to a previously unknown fault line, so was New Zealand, and as we can read, in part, as further confirmed by London's Independent News Service, "The devastating earthquake that tore through Christchurch on Tuesday is the product of *a new fault line* in the Earth's crust that seismologists were previously unaware of."[34] [Emphasis added.]

Was opening up a new fault line the objective of the earthquake? In Chapter 4, we looked at connections between Big Oil, fault lines, and hydrocarbons as per the Gulf of Mexico and New Madrid Seismic Zone. Is something similar going on off the coast of New Zealand? Geo-terrorism researcher Jeff Phillips thinks so. In fact, he feels that Anadarko Petroleum Corporation is HAARP's primary beneficiary.

> The company's portfolio of assets encompasses premier positions in the Rocky Mountains region, the southern United States and the Appalachian Basin. The company also is a premier deepwater producer in the Gulf of Mexico, and has production in Alaska, Algeria and Ghana with additional exploration opportunities in West Africa, Mozambique, Kenya, South Africa, Colombia, Guyana, New Zealand and China.[35]

Anadarko's board of directors is interesting: CEO James Hackett, a director of Halliburton, member of Trilateral Commission and Bilderberg; Luke Corbett, director of Kerr-McGee (owned by Anadarko); Peter Geren, former Secretary of both Army and Air Force; Paula Reynolds, director of BAE Systems Advanced Technologies that built HAARP's phase 2 upgrade (completed in 2006) and owns the patent (US7218571) for a magnetically driven underwater pulse generator; and General Kevin Chilton, former NASA astronaut and head of U.S. Space Command after his command at STRATCOM.

Most convincing, however, is that Anadarko's man, U.S. Representative Kevin Brady (R-TX), was in New Zealand Prime Minister John Key's office when the Christchurch quake occurred; four hours later, Anadarko had its permit for offshore exploration, for which it will use its ultra-deepwater drillship.[36]

Whereas Italy sentenced six seismologists and a FEMA-like official to six years in prison for manslaughter after reassuring the public that the M6.3 earthquake

---

34 "US Earthquake Weapon Test Fails Again."

35 www.anadarko.com/About/Pages/Overview.aspx

36 Thanks to Northland New Zealand Chemtrails Watch. See Geo-Terrorism, geo-terrorism.blogspot.com/2011/03/geo-terrorism-and-technetronic-warfare.html, and don't miss Jeff Phillips' grand piano improvisation of "Tornadoes."

that struck L'Aquila on 6 April 2009 would not happen,[37] in Christchurch, the Canterbury Earthquake Recovery Act (CERA) was passed and the Minister for Earthquake Recovery issued a Notice of Intention to Take Land:

> . . .negotiations for voluntary acquisition may run until the end of December [2011]. If agreement has not been reached by then, CERA's Christchurch Central Development Unit (CCDU) intends to take the land under compulsory acquisition [for public works].[38]

Claims for compensation are always *possible* but uncertain at best. Headlines like "Poverty 'rampant' in quake-hit Christchurch" are to be expected,[39] while corporations like Anadarko reap fortunes from the devastation.

Project Seal, a joint U.S.-New Zealand test carried out around New Caledonia and Auckland during World War II, may have been looking for new fault lines: 3,700 bombs were exploded in New Caledonia and later at Whangaparaoa Peninsula near Auckland, which created a 33-foot tsunami capable of inundating a small city.[40] This segues to three major 2011–2012 quakes and one tsunami in Japan that inundated the entire planet with radiation.

Japan

The September 2009 landslide victory of Japanese Prime Minister Yukio Hatoyama did not please Washington, D.C., his victory having been due largely to a pledge to the Japanese people to close the U.S. Marine Corps Futenma Air Station on Okinawa. Hatoyama wanted an Asia-centric future and his first steps included withdrawing Japanese naval forces in the Indian Ocean providing non-combat support for U.S. troops in Afghanistan and announcing that Japan would back a US$5 billion aid plan for Afghanistan,[41] plus visiting Nanjing (Nanking) for the 75th anniversary of the 1937 massacre of Chinese citizens under Japanese occupation, his purpose being to formally apologize.

To say the least, an Asian version of the European Union was not in Washington, D.C.'s plans. In November 2009, U.S. Defense Secretary Robert Gates warned Japan of "serious consequences" if U.S.-Japan commitments were not honored.[42] The day after Gates' comment appeared in *The Malaysian*, a M7.0 earthquake struck off the coast of Japan. The following April and May, thousands of Japanese protestors

---

37 Laura Margottini and Michael Marshall, "Bugged phone deepens controversy over Italian quake." *New Scientist*, 29 October 2012. The official who three days after the quake ordered seismologist Enzo Boschi to conceal foreknowledge, however, did *not* go to prison (*La Repubblica*).

38 Christchurch lawyer Margo Perpick, "The Lay of the Land." *The Press*, 27 November 2012.

39 Fairfax, January 29, 2013.

40 Jonathan Pearlman (Sydney), "'Tsunami bomb' tested off New Zealand coast." *The Telegraph*, 01 January 2013.

41 John Cherian, "Rethink of Relations." *Frontline* (India's national magazine), February 27–March 12, 2010; "Japan PM apologises for breaking US base vow," Kantipur.com, May 23, 2010.

42 "Japan Rethinks US Relationship." *The Malaysian*, February 25, 2010.

demonstrated against the American base, and in June 2010, Hatoyama tendered his resignation after only nine months in office. Four months later, October 18–29, 2010, Japan hosted the Convention on Biological Diversity attended by 193 nations. The U.S. refused to ratify a moratorium that stated:

> that no climate-related geoengineering activities that may affect biodiversity take place, until there is an adequate scientific basis on which to justify such activities and appropriate consideration of the associated risks for the environment and biodiversity and associated social, economic and cultural impacts, with the exception of small-scale scientific research studies. . .

The ink on the moratorium was barely dry when the U.S. House Science and Technology Committee issued the first congressional report on geoengineering and rubberstamped "research," "experiments," and "tests" under the NSF and NOAA.[43]

A few months short of the 66 years since the U.S. had unleashed the previous nuclear hell on Japan, an all too similar hell broke loose:

I. *M9.03 Tōhoku undersea megathrust earthquake, 11 March 2011* (20,000 dead). This most powerful earthquake to ever hit Japan moved the main island of Honshu 2.4 m. (8 ft.) east, shifted the Earth on its axis 1–25 cm (4–10 in),[44] and sent 133-foot tsunami waves six miles inland, causing level 7 meltdowns at three reactors in the Fukushima Daiichi Nuclear Power Plant.

II. *M7.1 Fukushima Hamadōri earthquake, an "intraplate aftershock," 11 April 2011* (four dead).

III. *M7.3 earthquake east of Sendai, Japan, December 7, 2012.* Also in the "aftershock zone" of the March 2011 Tōhoku earthquake."[45] Note the date: 71 years to the day after Pearl Harbor.

From March 12 to May 4, 2011, Operation Tomodachi—a "humanitarian mission" of 70,000 U.S. military men and women from the Navy, Air Force, Army, and Marines—delivered food, blankets, and water to devastated areas.

"Aftershock" is a misleading term. Each aftershock is actually an earthquake of any size and location that would not have taken place, had the main shock not struck. Many M4.5s and greater occurred as far away as Mexico and Japan after the M8.6 East Indian Ocean strike-slip earthquake on April 11, 2012, a year after

---

43 Juliet Eilperin, "Geoengineering sparks international ban, first-ever congressional report." *Washington Post*, October 29, 2010.

44 Kenneth Chang, "Quake Moves Japan Closer to U.S. and Alters Earth's Spin." *New York Times*, March 16, 2011.

45 "Aftershock: Fukushima Japan Just Rocked With Another Devastating Earthquake." *Before It's News*, December 7, 2012.

the Fukushima quakes. *The San Andreas fault line is also a strike-slip.*[46] Despite the high Richter scale readings, the U.S. Geological Survey chose to categorize the last two Japanese quakes as aftershocks, of which there were more than 5,000. (Haiti showed 42 high-magnitude aftershock entries between January 12 and 14.[47])

The first T hoku quake surprised seismologists. As *Wired* Science News put it, "The epicenter of the quake was about 80 miles east of the city of Sendai, in a strip of ocean crust previously thought unlikely to be capable of unleashing such energy."[48] Freelance journalist Jim Stone (quoted in Chapter 3 regarding JP-4 jet fuel) uncovered a multitude of nonsequiturs regarding the M9.03 story, such as that the only significantly damaged areas were those hit by the tsunami, including the Fukushima nuclear plant, and that the size of the earthquake that hit off Japan's coast actually registered M6.67, while the tsunami was what would have occurred during an actual M9.03.

And what caused the massive explosions in Reactors 3 and 4? Nuclear weapons, Stone avers. Prior to the disasters, Magna BSP, a security firm, installed massive "security cameras" inside the reactors—security cameras weighing more than 1,000 pounds and looking like uranium gun-type nuclear bombs. Stone claims that Japan was planning to enrich uranium for Iran.[49]

Two days before the third quake on December 7, the U.S. conducted yet another nuclear "experiment" at the Nevada National Security Site:

> The Pollux subcritical experiment was carried out by scientists at the Los Alamos, New Mexico national laboratory and the Sandia National Laboratories and involved a tiny sample of plutonium bomb material. Subcritical nuclear experiments have been conducted in the U.S. since 1997 in order to help scientists understand how plutonium ages in the stockpile. They use chemical explosives to blow up bits of nuclear materials designed to stop just short of erupting into a nuclear chain reaction, also known as a criticality . . .International inspectors were not allowed to witness the experiment, as Washington has prevented access to its test site since the late 1990s. Wednesday's test is the 27th American "*subcritical experiment*" since full-scale nuclear weapons tests were halted in 1992.[50]

Did the U.S. use a nuclear bomb off Japan's east coast on Pearl Harbor Day, after Iran and Japan condemned the Nevada test and after leaders of Hiroshima and Nagasaki spoke out, saying that the test was to prove that the U.S. "could use nuclear weapons anytime"[51]?

---

46 "Rare Great Earthquake in April Triggers Large Aftershocks All Over the Globe." *TerraDaily*, September 27, 2012.

47 earthquake.usgs.gov/earthquakes/recenteqsww/Quakes/quakes_all.php

48 Devin Powell, "Japan Quake Epicenter Was in Unexpected Location." *Wired*, March 17, 2011.

49 Environmental Terrorism, jimstonefreelance.com/fukushima.html

50 "US nuclear test condemned by Iran, Japan." RT.com, 8 December 2012.

51 Ibid.

On his *UrbanSurvival* blog site on March 16, 2011, financial wizard and shortwave aficionado George Ure studied public record sources and made several acute observations along HAARP lines, which I will quote at length:

At least, insofar as the magnetometer readings about the government/University of Alaska HAARP, which, as you may recall from our earlier discussions, uses a huge steerable HF radio transmitter system which is powered with a roughly 2-megawatt power plant which, given the size of their antenna array, pumps many megawatts (if not gigawatts) of effective radiated power (ERP) into the ionosphere for "research."

So powerful is the radiation off the HAARP site that they even have their own radar coverage so that aircraft in the area aren't inadvertently damaged. In other words, lotsa, lotsa power.

. . .it's always been *conjecture based on the magnetometers* since I've been unable to locate the critical data needed to interpret whether HAARP was *causative* to some of the odd phenomena afoot in the world today, like bird kills, out-of-place earthquakes, and the like, or whether it was *coincident* to anomalous events.

The data which would be required in order to make such an assessment necessarily include: Effective radiated power (ERP), the directional heading of the array — since in high frequency (HF) radio, there are always two paths to anywhere on earth (the short path and the long path which is the reciprocal direction from the antenna heading) as well as take-off angles, since the take-off angle of a radio wave may be varied.

. . .Since the public doesn't get access to transmitter logs with frequency, pulse type, power, direction, and take-off angles, the only thing we know is that yes, HAARP's magnetometers were on and working at the time of the 9.0 quake off Honshu, but follow me here: this is where things get interesting:

I plotted out the times of the Japan quake plus a few other quakes (6.0 and above) that were temporally adjacent, and put a few of them on the magnetometer timeline from HAARP's database and there is a big spiky thing around 07:00=08:00 UTC on March 10. Guess what's there? A 4.8 off Honshu . . .That's the big black spike which is followed by the next big black spike smack dab coincident with the 9.09 quake! The 6.1 off Honshu didn't do anything, yet a 4.8 is coincident and then the 9.0? Like wait for something, then pump it maybe? Of course not! Why, such thinking would verge on conspiratorial, would it not? But then again, there's the data to mull . . .and certainly a mechanism *might be postulated* . . .

If you want to have a ton of fun, you could go into the magnetometer chain and look up the data for the magnetometer chains in the general direction of Japan, and what do you find? Missing data for places like Trapper Creek and

Anchorage which, although they are in the general (short path) direction of Japan, is only coincidental . . .we're sure . . .try to fake a look of surprise right about here . . .

. . .Japan as you probably know has a *huge unfunded pension liability.* Nearly a quarter of a trillion dollars in 2009 and still growing, according to this Bloomberg report from January of 2010: "Japan's top 278 companies were a combined 21.5 trillion yen ($235.7 billion) behind on their pension funding in fiscal 2009."[52] A fair-minded business reporter would be looking — about now — for the [Powers That Be] to tip their hand on what happens to those pension liabilities . . .national emergency — that kind of thing.

I don't know about you, but I'd sure like to see HAARP open up and report all of its transmitter operations and array headings for the week prior to the Japan quake. Not that I'm asserting any wrongdoing, of course. BUT I am bothered down at the soul level by what jumps out of the magnetometer readings.

That all earthquakes are not equal.[53]

No, not all earthquakes are created equal. A magnetometer—which can be used to predict earthquakes as well as give evidence of having created one—registers disturbances in the magnetic field in the Earth's upper atmosphere, whereas a seismometer measures activity in the ground. The HAARP magnetometer data at its website (now shut down) indicated that HAARP began broadcasting the earthquake-inducing frequency of 2.5 Hz into the ionosphere on March 8, 2011 and continued through March 9 and March 10, not turning it off until ten hours *after* the M9.0 quake on March 11. By beaming the frequency at a specific trajectory, HAARP can trigger an earthquake anywhere on Earth.[54]

[HAARP's virtual ionospheric] ELF antenna can emit waves penetrating as deeply as several kilometers into the ground, depending on the geological makeup and subsurface water condition of a targeted area. HAARP uses ground penetrating radar (GPR) to beam pulses of polarized high-frequency radio waves into the ionosphere. These pulses can be fine-tuned and adjusted so that the bounced ground penetrating beam can target a very specific area and for a specific length of time. Beaming a very large energy beam in the ground for an extended period can cause an earthquake. After all, an earthquake is the result of a sudden release of energy in the Earth's crust that creates seismic wave resonances. The same array of antennas that are used by HAARP for

52 Jason Clenfield and Tomoko Yamazaki, "JAL Pension Shortfall May Prompt Japan Inc. to Change Its Ways." *Bloomberg,* January 21, 2010.

53 Entire article (minus the chart with the "big spiky things") available at stienster.blogspot.com/2011/03/george-ure-haarp-data.html

54 "US Government Takes Down HAARP Website to Conceal Evidence of US Weather Modification and Earthquake Inducing Warfare." *Community Chemtrail News,* May 1, 2011.

> ground penetrating tomography can also be used to penetrate deep into the ground and create an earthquake anywhere around the world.[55]

The May 18, 2011 article in *MIT Technology Review*, "Atmosphere Above Japan Heated Rapidly Before M9 Earthquake," also supports Ure's contention. The Physics arXIV team had looked at "some fascinating data from the DEMETER spacecraft showing a significant increase in ultra-low frequency radio signals before the magnitude 7 Haiti earthquake in January 2010." Dimitar Ouzounov at the NASA Goddard Space Flight Center in Maryland:

> . . .before the M9 earthquake, the total electron content of the ionosphere increased dramatically over the epicentre, reaching a maximum three days before the quake struck. At the same time, satellite observations showed a big increase in infrared emissions from above the epicentre, which peaked in the hours before the quake. In other words, the atmosphere was heating up.[56]

Similarly, a piezoelectric[57] explanation was trotted out to explain Freund's observation of the "huge" signal in the ionosphere before the 12 May 2008 M7.8 quake in China. Freund and his father Friedemann Freund, also working at NASA's Ames Research Center, even invented a new term to account for the increase in infrared emissions coming from the ionosphere:

> According to their theory, the charge carrier is a "positive hole" known as a phole, which can travel large distances in laboratory experiments. When they travel to the surface of the Earth, the surface becomes positively charged. And this charge can be strong enough to affect the ionosphere, causing the disturbances documented by satellites. When these pholes "recombine" at the surface of the Earth, they enter an excited state. They subsequently "de-excite" and emit mid-infrared light particles, or photons. This may explain the IR observations.[58]

After contemplating MIT's infrared emissions chart, *qbit* responds:

> Granted, it's plausible that piezoelectric discharges in the rock during an earthquake could naturally generate atmospheric plasma, but this doesn't account for the IR [infrared] emissions. Especially notable is the circular mode of the IR signal, whereas one might expect to see an irregular pattern

---

55 "Link between HAARP earth-penetrating tomography technology and earthquakes." *World News*, Sunday, March 13, 2011.

56 "Atmosphere Above Japan Heated Rapidly Before M9 Earthquake." *MIT Technology Review*, May 18, 2011.

57 *Piezoelectric* refers to rocks acting like batteries when they are compressed by tectonic plate shifts.

58 Paul Rincon, "Plan for quake 'warning system.'" *BBC News*, 5 June 2008.

> if the IR emissions came from a geological fault. A directed energy weapon like HAARP or a Free Electron MASER could also provide an explanation for the atmospheric phenomena, but without more information it's impossible to infer a causal relationship.[59]

"Pholes" seems like a thin deflection from ionospheric heater activity, as does MIT's Lithosphere-Atmosphere-Ionosphere Coupling Mechanism theory, MIT being one of HAARP's many university collaborators (University of Alaska, Stanford, Penn State, Boston College, Dartmouth, Cornell, University of Maryland, University of Massachusetts, Polytechnic University, UCLA, Clemson University, and the University of Tulsa), and NASA Ames being—well, NASA.

On December 15, 2012, the *Japan Times* reported a Tohoku University study regarding the possibility of a M10 earthquake occurring and how the tsunami from such a quake would continue for several days and cause damage to other Pacific Rim nations. Such an earthquake could only occur once every 10,000 years, the professor at Tohoku University's Research Center for prediction of Earthquakes and Volcanic Eruptions promised.[60]

Was this "study" a way of subtly warning the world without saying anything outright? On February 2, 2013 colored lights "fell" on Hokkaido right before a M6.4 hit. Two months later on April 13 a M6.3 hit western Japan, and the next day a M5.2 hit near Fukushima, the epicenter at a depth of 50 km off the Pacific coast of Japan. More than 390 aftershocks were recorded.

HURRICANES

Hurican is the name of the Carib god of evil. To Westerners, a hurricane occurs when an intense low pressure is surrounded by a violent rotating storm whose wind speed reaches at least 75 miles per hour. It's a hurricane in the North Atlantic, the Northeast Pacific east of the dateline, and the South Pacific east of 160E, but west of the dateline it's a typhoon, and in the Indian Ocean, a cyclone. The Saffir-Simpson scale formulated in 1969 posits five categories:

Category 1: 75–95 mph winds, storm surge up to 1.5 meters
Category 2: 96–110 mph winds, storm surge 1.8–2.4m.
Category 3: 111–130 mph winds, storm surge 2.7–3.7m.
Category 4: 131–155 mph winds, storm surge 4–5.5m.
Category 5: +155 mph winds, storm surge +5.5m.

59 chemtrails.cc/tags/haarp/. YouTube shows a glowing blue plasma ball near the horizon as the quake begins—definitely not a piezoelectric phenomenon, nor ball lightning.

60 "Magnitude 10 temblor could happen: study." *Japan Times*, December 15, 2012.

In 2004 alone, hurricanes struck Florida once a week for a month. As Michael McCarthy of the UK's *Independent* put it, "To have three tempests of the intensity of hurricanes Charley, Francis, and Ivan burst through one state in the space of four weeks is certainly unusual."[61]

> *Category 3–4 Hurricane Charley*, east Gulf Coast, Saturday, August 14; 2 million evacuated, 27 dead, $7–8 billion damages.
>
> *Category 2–3 Hurricane Francis*, eastern Atlantic Coast, Saturday, September 4; 2.4 million evacuated, 0 dead, $2–4 billion damages.
>
> *Category 4–5 Hurricane Ivan*, 10-day Caribbean tour, September 5–September 14; 119 dead, $18 billion damages.
>
> *Category 2–3 Hurricane Jeanne*, from Haiti to New Jersey, September 16–28; 3,006 dead in Haiti, 7 in Puerto Rico, 18 in Dominican Republic, 4 in Florida, damages in U.S. alone $6.9 billion.

Were these four hurricanes close-to-home "tests" of the HAARP- chemtrails pump-and-dump so the international community would believe they were normal seasonal catastrophes meted out by Nature? For some, the aftermaths were devastating; for the players discussed in Chapter 6, quite profitable.

The "Hurricane Intervention" chart [see page 125] from the *Scientific American* cover article (October 2004), "Controlling Hurricanes: Can Hurricanes and Other Severe Tropical Storms Be Moderated or Deflected?" by Dr. Ross N. Hoffman, is instructive, especially given that HAARP, with its at that time unique steering capability, was up and running in 2004, though of course Hoffman does not mention it. Instead, he comments (in the usual subjunctive mood) that "tiny influences" coupled with a steering capability would be a powerful combination:

> A chaotic system is one that appears to behave randomly but is, in fact, governed by rules. It is also highly sensitive to initial conditions, so that seemingly insignificant, arbitrary inputs can have profound effects that lead quickly to unpredictable consequences. In the case of hurricanes, small changes in such features as the ocean's temperature, the location of the large-scale wind currents (which drive the storms' movements), or even the shape of the rain clouds spinning around the eye can strongly influence a hurricane's potential path and power.
>
> The atmosphere's great sensitivity to tiny influences—and the rapid compounding of small errors in weather-forecasting models—is what makes

61 Michael McCarthy, "Four Hurricanes In 5 Weeks: What, Exactly, Is Going On?" *The Independent*, 21 September 2004.

> long-range forecasting (more than five days in advance) so difficult. But this sensitivity also made me wonder *whether slight, purposely applied inputs to a hurricane might generate powerful effects that could influence the storms,* whether by steering them away from population centers or by reducing their wind speeds.[62] [Emphasis added.]

Hoffman echoes geophysicist Dr. Gordon J.F. MacDonald's statement, "The key to geophysical warfare is the identification of environmental instabilities to which the addition of a small amount of energy would release vastly greater amounts of energy,"[63] to which *Weather Warfare* adds, "Clearly if one can steer a hurricane away from a population center, one can, using the exact same technology, steer it into one."[64]

In September 2005, Hoffman appeared in the Swiss documentary *Power Over the Weather*:

> [Transcript:] Meteorologist Ross Hoffman works at the Institute for Atmospheric Research in Massachusetts. *He steers hurricanes. . .*
>
> [Hoffman:] We looked more closely at Hurricane Iniki, which raged in Hawaii in 1992, and Hurricane Andrew, which devastated Florida [also in 1992]. In our first experiment, we tried to simulate these storms and predict their development . . .Then we look for conditions which might weaken the storm or alter its position, and which are highly sensitive in their reaction to the smallest changes. Situations in which small changes can later have major effects.[65]

In disaster capitalism terms, earthquakes and hurricanes create new real estate and reconstruction deals, whether it's Christchurch in New Zealand or the two costliest American hurricanes ever, Katrina (August 29, 2005) and Superstorm Sandy (October 29, 2012). To generals and geoengineers, that people died and lost their homes and livelihoods is simply the collateral damage expected of full spectrum dominance.

Hurricane Katrina and the Port of New Orleans

Hurricane Katrina's size and force were gargantuan; its eye alone was the size of Lake Pontchartrain. To understand this hurricane, we must begin with politics in the age of disaster capitalism, especially now that the HAARP- chemtrails

62 Ross N. Hoffman, "Controlling Hurricanes: Can Hurricanes and Other Severe Tropical Storms Be Moderated or Deflected?" *Scientific American*, October 2004.

63 Gordon J.F. MacDonald, "Geophysical Warfare: How to Wreck the Environment." *Unless Peace Comes: A Scientific Forecast of New Weapons*. Ed. Nigel Calder. A. Lane, 1968.

64 Jerry Smith, *Weather Warfare: The Military's Plan To Draft Mother Nature*. Adventures Unlimited Press, 2006.

65 From *Symptomatologische Illustrationen*, Lochmann-Verlag. Translated by Graham Rickett, a friend.

pump-and-dump provides the global muscle. Here is *Veterans Today* columnist Trowbridge Ford:

> When the December [2004] tsunamis had the desired effect of bringing the international community back on line . . .it was hardly surprising that the Bush administration looked to the NRO [National Reconnaissance Office[66]] again to cook up more clouds when it came to bringing the increasingly dubious American public back on line, especially with the anniversary of the 9/11 attacks rapidly approaching. [Hurricane Katrina occurred thirteen days before the 4th anniversary of 9/11.]
>
> The only trouble was that the planners had little to work with—with only the puny Tropical Depression Katrina which was slowly drifting across the Atlantic from Africa by mid-August. Quite possibly, it was increased in strength by Air Force using remotely controlled nanoparticles to seed and control cloud charge—sparking thunderbolts on command. By the time it reached Florida, it was still only a Category 1 hurricane, as Michael Moore duly recorded afterwards in an angry letter to President Bush: "Last Thursday, I was in south Florida while the eye of Hurricane Katrina passed over my head. It was only a Category 1 then but it was pretty nasty. Eleven people died. . ." As the report "Weather As A Force Multiplier: Owning the Weather in 2025"—put together by a group of military experts in 1996—said: "The technology is there, waiting for us to pull it all together."[67]

Ford then describes how *a shot of energy* quickly turned the Category 1 into a Category 5 with water leading the wind instead of vice versa, waves at 60 feet, and coastal water temperature at 90°F (32°C). Oil rigs were ripped from their moorings and driven miles away, cutting production by 2 million barrels per day and thus driving oil prices (and windfall profits for oil companies like Anadarko Oil) up, up, and up.

If you perused the Web early enough during Katrina's assault, you may have caught the Doppler space footage (before it was removed) of two titan energy vortices battling each other at the mouth of the Mississippi River. Easterly Katrina was turned from her dead-on course by a *counter-force* from the west—a tussle that may have necessitated a Plan B, namely the one-by-one detonation of levees the next day—like Building 7 of the World Trade Center in New York City on September 11, 2001. Former KPVI-TV weatherman Scott Stevens blamed "a Russian-made electromagnetic generator" (the Woodpecker?) for Hurricanes

---

66 Created in 1960, the "black" NRO did not go public until 1992. Wikipedia: "one of the 'big five' U.S. Intelligence agencies. [The NRO] designs, builds, and operates the spy satellites of the United States government, and coordinates the analysis of aerial surveillance and satellite imagery from several intelligence and military agencies."

67 Trowbridge Ford, "Glimpses of America's Man-Made Disasters, Part 5." cryptome.org, May 24, 2009.

Ivan and Katrina, in part because of their Russian names. Was a Russian ionospheric heater one of the titans vying for power over New Orleans' fate? Stevens' theory was called "ludicrous" and the second law of thermodynamics invoked (that energy can neither be created nor destroyed) to cover up invisible HAARP and plasma technology.[68]

Ford referred to the political aftermath of Katrina as an unraveling "cock-up." Land and cell phone systems collapsed so Clear Channel radio and TV could jam shortwave frequencies and monopolize coverage so that looters could be overemphasized, lives lost minimized, and suspicious levee breaches and pump failures ignored. Aaron Broussard, president of Jefferson Parish, said on *Meet the Press* that FEMA even cut the lines of the parish's emergency communications system. The National Guard turned the Red Cross away, and FEMA and the federal government stood watching *for five days* before doing much of anything.

As with 9/11, the questions left are virtually endless. Robert Schoen, a resident of New Orleans, asked:

> Why did breaches in the levees occur only many hours after the hurricane had already passed? Why was there a failure of all the city's pumps when they were needed most? How was the media able to reorganize so quickly when so much of the city's government and vital infrastructure left disorganized, without communication and in tatters for days on end? Why were ham radio frequencies being jammed? Why was there an endemic failure of the cell phone towers, each of which had their own generator? Why weren't emergency temporary mobile towers and other alternative forms of communication brought in immediately, if only for city management use?
>
> *Why were dogs in the city running around in circles and acting so wired before the hurricane's approach? Dog owners in the city noted that their dogs' actions before this hurricane were different than from others. Perhaps the dogs were feeling a large scalar signal pointed at New Orleans. The HAARP array in Alaska is one such technology capable of weather control.*[69] [Emphases added.]

Two days after Katrina, the thought-provoking article "New Orleans: A Geopolitical Prize" by STRATFOR founder George Friedman appeared at STRATFOR's Internet site.[70] Called "a private quasi-CIA" by *Barron's* financial magazine, STRATFOR is based in Austin, Texas. Indeed, Friedman's oblique references seem to sprout from inside knowledge as to what "geopolitically" occurred at the Gulf of Mexico port city. Friedman begins, "Until last Sunday [two days previous], New Orleans was, in many ways, the pivot of the American economy," then goes on to compare the devastation to the Battle of New Orleans

---

68 "Cold-war device used to cause Katrina?" *USA Today*, September 20, 2005.

69 Robert Schoen, "Katrina: Natural Disaster or Sabotage and Institutional Terrorism?" 7 September 2005. Strangely, this version has disappeared from the Web.

70 George Friedman, "New Orleans: A Geopolitical Prize." STRATFOR, 01 September 2005.

in January 1815, the German U-boats at the mouth of the Mississippi in World War II, and the purported Cold War fear: *If the Soviets could destroy one city with a large nuclear device, what would it be?*

He then blames "nature" for the assault, though his Manhattan Project choice of imagery and undertone allude again to (space-age) weapons:

> Last Sunday, nature took out New Orleans almost as surely as a nuclear strike. Hurricane Katrina's geopolitical effect was not, in many ways, distinguishable from a mushroom cloud.

He goes on to bemoan how now-homeless laborers and craftsmen once central to the vital harbor must now seek new homes and livelihoods. Then again, he references war:

> It is in this sense, then, that it seems almost as if a nuclear weapon went off in New Orleans. The people mostly have fled rather than died, but they are gone . . .The area *can* recover, to be sure, but only with the commitment of massive resources from outside—and those resources would always be at risk to another Katrina . . .Protecting that port has been, from the time of the Louisiana Purchase, a fundamental national security issue for the United States. . .

Friedman moves from a paean to New Orleans to a creed of faith in geopolitics, whatever the cost, hinting again that "nature" is now a geopolitical weapon:

> Geopolitics is the stuff of permanent geographical realities and the way they interact with political life. Geopolitics created New Orleans. Geopolitics caused American presidents to obsess over its safety. And geopolitics will force the city's resurrection, even if it is in the worst imaginable place.

Three weeks after Katrina on September 20, 2005, Category 5 Hurricane Rita—the fifth major hurricane of the 2005 season—passed through the Florida Straits, hit the abnormally warm waters of the Gulf, and made a beeline for Texas' Gulf coastline, carefully circumventing New Orleans, Houston, and Galveston. Forecasters, the National Reconnaissance Office (NRO), military operators, and rescue services were better prepared this time: 2.5–3.7 million people were evacuated.

Perhaps after studying New Orleans and seeing the writing on the wall for southern nations, former Malaysian Prime Minister Mahathir Mohamed announced that he would speak at the SUHAKAM conference in Kuala Lumpur in conjunction with Malaysian Human Rights Day on September 9, 2006. SUHAKAM is the Human Rights Commission of Malaysia and an A-status member of the Asia Pacific Forum (APF). Mahathir revealed how back in 1997 the *Wall Street Journal*[71] had lied about the *steerable* technology used to help Malaysia clear its smog (and capable of creating cyclones) being Russian. (See Chapter 7 on Igor Pirogoff's Elate Intelligent Tech.) The steerable technology had come from Raytheon's E-Systems, owner of HAARP patents, with the approval of Pentagon chief William Cohen.

Ford then writes, "To silence Mahathir, Britain's High Commissioner to Malaysia Bruce Cleghorn, the country's chief of state, American Ambassador Christopher LaFleur, and several other diplomats walked out of the conference before Mahathir got the chance to open his mouth."[72]

Hurricane Sandy and the Port of New York and New Jersey

Between 2005 and 2012, many hurricane-level storms rose and fell, but Hurricane Sandy or "Frankenstorm" was voted by mass media as the "storm of the century" that left thousands homeless and required $50.7 billion from Congress.[73] Internet observers combed over articles[74], photographs, and YouTube clips as they struggled to comprehend the *viciousness* of what *Veterans Today* columnist Harold Saive called a cold-blooded "coordinated exercise":

> Like 9/11, Frankenstorm Sandy could more accurately be described as a coordinated exercise than an unplanned disaster. What emerges is a clear record of media, government, military and inter-agency secrecy and complicity not acknowledged since the Manhattan Project.[75]

Before landfall on October 29, 2012, nanoparticulate aerosols were dumped from 40,000 feet into the hurricane core. Satellite Doppler indicates two layers of clouds, one of natural thunderheads, the other chemtrail-created. Saive says that the hurricane was manipulated under Operation HAMP (Hurricane Aerosol Microphysics Program) of the Department of Homeland Security (DHS). HAMP operates in tandem with the DHS WISDOM Project (Weather In-Situ Deployment Optimization Method), Unmanned Aerospace Systems (UAS) projects (drones), and NOAA aircraft missions.

A YouTube video accompanying Saive's comments presents PowerPoints and audio of a speech by Dr. Joe Golden at the 2010 American Meteorological Society (AMS) conference, "Identification and Testing of Hurricane Mitigation Hypotheses," an outgrowth of the February 2008 DHS/ESRL (Earth System Research Laboratory at NOAA) workshop in Boulder. Golden is an elected

71 Chen May Yee, "Malaysia to Battle Smog with Cyclones." *Wall Street Journal*, November 13, 1997.

72 Ford, "Glimpses of America's Man-Made Disasters, Part 5," 2009.

73 "Displaced Sandy Victims: Temp Housing Unlivable." *CBS News*, January 28, 2013. "Thousands still homeless after Sandy." RT.com, April 29, 2013.

74 For example, Jim Lee, "Geoengineering Frankenstorm: Hurricane Sandy and the Air Force Weather Weapon System," Parts 1 and 2, November 2, 2012.

75 Harold Saive, "HURRICANE SANDY: The HAARP and Aerosol Geoengineered Storm of Century?" examiner.com, October 30, 2012.

fellow of the AMS, member of the American Geological Union, the Royal Meteorological Society, Sigma Xi, and on the board of directors of the American Association for Wind Engineering (AAWE) in Fort Collins, Colorado. After stressing that "Aerosols affect cloud micro-physics and dynamics in a very fundamental, important way," Golden talks up how HAMP seeks to test "mitigation hypotheses by means of rigorous numerical simulations supported by necessary observations" by utilizing small CCN (cloud condensation nuclei) aerosols and radiation-absorbing aerosols. HAMP's objectives include:

- Increase updraft velocities
- Change drop-size distributions
- Transform tropical clouds into thunderstorms
- Affect the intensities and tracks of hurricanes

Golden admits that aerosols can also *induce* floods and droughts, then stutters that yes, aerosols were used to manipulate Katrina's intensity.

Katrina and Sandy turned both ports of New Orleans and New York-New Jersey into "real estate deals." Both harbors and the neighborhoods around them are now being redesigned and reassigned. Katrina may have been a Gulf of Mexico basin experiment or "test" and Sandy an Atlantic Ocean basin experiment or "test" under NASA's Hurricane and Severe Storm Sentinel (HS3), the purpose being to address "the controversial role of the Saharan Air Layer (SAL) in tropical storm formation and intensification as well as the role of deep convection in the inner-core region of storms" needing "sustained measurements over several years."[76] More hurricanes mean a larger database. High above the debris of ravaged human lives, NASA Global Hawk UASs (unmanned aircraft systems) observe storm formation for up to 30 hours. I'm sure the view from 55,000 feet, unlike the view on the ground, is spectacular *and distant.*

TORNADOES

In season 1, episode 2 of the Weather Channel's *Hacking the Planet* series, Swiss inventor Slobodan Tepic (doctorate from MIT) was interviewed about his Atmospheric Vortex Engine (AVE) that harnesses the energy of *microwave-induced* tornadoes. Like Tesla, Dr. Tepic hoped his invention would help humanity, but of course the deep pockets that seek to weaponize everything on Earth have other ideas:

> The [Canadian] company, AVEtec Energy Corporation, is working on a proof-of-concept device after winning this month a grant from The Thiel Foundation,

76 espo.nasa.gov/missions/hs3

> a non-profit organization created by PayPal co-founder Peter Thiel. The [AVE] is meant to make a rotating column of air that sucks air from its base like a natural twister. The company already built a four-meter diameter prototype and will now use the grant money to construct a version twice as big, which it hopes will produce a 40-meter-tall vortex with a diameter of 30 centimeters . . .It can potentially go up as far as 20 kilometers for a vortex with the base diameter of 100 meters generated in a 200-meter diameter 80-meter-high tower, the company's website explains.[77]

And of course, the AVE will need field-testing during tornado season. . .

Between May 4 and 10, 2003, nearly 400 tornadoes occurred in 19 states—42 deaths and billions of dollars in damage. The sheer number of tornadoes that have occurred in central and eastern Oklahoma since the HAARP-chemtrails pump-and-dump went fully operational is concerning: October 4, 1998 (28); May 3, 1999 (60); May 10, 2010 (35). The little town of Moore, smack dab in the Tornado Alley path, was consistently hit.

Between May 15 and 17, 2013, an "outbreak" of 23 tornadoes swept through northern Texas, south-central Oklahoma, northern Louisiana and Alabama, including an EF-4[78] in Granbury, Texas (six people dead) and an EF-3 in Cleburne, Texas. On May 20, 2013, an EF-5—the strongest category—mowed a 1.3-mile swath through Moore and Newcastle, Oklahoma: devastation for 40 minutes and 17 miles[79], winds of 210 miles per hour ravaged 13,000 homes (many built by Habitat for Humanity), with damages quoted at $2 billion. Thirty-three thousand people were affected, 24 dead, hundreds injured.

For steering weather locally, NEXRAD (Next Generation Weather Radar) WSR-88D (Weather Surveillance Radar, 1988, Doppler) stations are handy, with their evenly spaced transmitters and receivers. Besides the old "decommissioned" GWEN towers that have been updated and recommissioned, and the ubiquitous cell towers, radio stations like those in Moore's vicinity (KSGF, KEAX, KVNX, KICT, KTWX, KSRX, KTLX, and KINX) often have Doppler and NEXRAD radar for up-to-date weather that double for national security "dual use."[80] NEXRADs run on 50.8 kW while their *klystron* (HPM electron beam tube) converts standard power to 750 kW of coherent energy that is then transmitted as pulse rotating frequencies at varying angles (elevations).[81]

To create the *wind shear* for dozens of simultaneous tornadoes or super-cells, NEXRAD pulsed rotation frequencies need an artificial precipitation thick with

77 "Man-made tornadoes may produce green energy." *RT*, December 27, 2012.

78 Enhanced Fujita Scale (EF Scale): the U.S. rating system for tornado strength.

79 "The Tornado Outbreak of May 20, 2013," www.srh.noaa.gov/oun/?n=events-20130520.

80 Dual use refers to "any technology which can satisfy more than one goal at any given time. Thus, expensive technologies which would otherwise only serve military purposes can also be used to benefit civilian commercial interests when not otherwise engaged, such as the Global Positioning System." — Wikipedia, "Dual-use technology"

81 See "Floods in Southern Alberta caused by NEXRAD Stations in the U.S. & Canada — SHOCKING EVIDENCE." *GeoEngineering Watch: Calgary Alberta*, June 21, 2013.

metal particulates and chaff. Once NEXRAD rotating frequency pulses strike the nanoparticulates and H_2O, a collision between HAARP-heated artificial precipitation (HOT) and frequency-activated ice nucleation (COLD) can be stirred into funnels. (Grapefruit-sized hail fell in Moore.) Then just steer it and feed it with more particulates.

Finally, plasma balls or orbs are often seen in the midst of tornadoes. These orbs are *not* UFOs; they are plasma space weapons directed by frequencies. In the case of tornadoes, they serve as guides for storm development, like an "eye of God." Plasma, laser, directed energy weapons (DEWs)—all can be magnified by frequency to affect weather for war or wealth, take out asteroids or meteors, even *dissociate molecules*.

White fire in Santa Barbara County's Los Padres National Forest during 2013 Memorial Day weekend while a M4.6 quake shook off the coast. Flooding in Norway and San Antonio, Texas. Termites swarm New Orleans. Fireballs over Cyprus and Ireland. Mt. Etna spewing and baffling Italian scientists. A tornado near a Russian nuclear plant. High over our heads, mirrors shimmer on low- and high-orbit satellites and cosmic chimera streak like augurs through the heavens as we begin our descent into even less visible schemes riding the HAARP-chemtrails pump-and-dump into planet Earth. Not only are we breathing in heavy metals and polymers while being inundated by an ionized atmosphere and radiation untold, but our blood has been drafted to play Petri dish for a nanotech hybrid "experiment" we have taken to calling Morgellons.

CHAPTER EIGHT

Morgellons: The Fibers We Breathe and Eat

To this day, few Americans know about the special top-secret program that brought German scientists to the United States after World War II, and fewer still know that their number included medical scientists. Code-named Operation Paper Clip . . .hundreds of 'specialists' . . .entered the United States under Joint Chiefs' protection, avoiding regular immigration procedures and requirements . . .It is hard to escape the conclusion that many of the German recruits were for decades important consultants on a myriad of military-medical projects.

— Jonathan D. Moreno, *Undue Risk: Secret State Experiments on Humans*, 1999

Surely, if the questions that should have been asked about the various applications of electromagnetic radiation continue to go unasked about future applications, this nation will one day be faced with a public health disaster of monumental and perhaps irreversible proportions.

— Paul Brodeur, *The Zapping of America: Microwaves, Their Deadly Risk, and the Coverup*, 1977

There is no such thing as a microorganism that cannot cause trouble. If you get the right concentration at the right place, at the right time, and in the right person, something is going to happen.

— George H. Connell, assistant to the director of the CDC; testimony before the U.S. Congress Senate Hearings before the Subcommittee on Health and Scientific Research of the Committee on Human Resources, *Biological Testing Involving Human Subjects By the Department of Defense*, March 8 and May 23, 1977

As strange black fibers pop up in McDonald's Chicken McNuggets,[1] and transgenic pigs, cows, and fish hit the supermarkets,[2] we move from the synergistic nightmare of endocrine-disrupting chemicals and acquired immunodeficiency syndrome (AIDS) to yet another siege on the global immune system: the ghastly condition known as Morgellons manifesting in some as suppurating sores with *erupting fibers that match the fibers dropping from chemtrails.*

The term *Morgellons* hearkens from the 17th century when eclectic author and alchemist Sir Thomas Browne (1605–1682) employed it for a disease that made its early home in Languedoc, France, where it was variously called *Les Crinons,* Masclous, and Masquelons. The *mouscouloun* was the hook attached to the end of a spindle, derived from the Latin *muscula,* meaning little fly—the attempt being to describe the bristles/insects/parasites besetting infants and small children that would only abandon their host when *coaxed* from beneath the skin by briskly rubbing the skin with honey.[3]

Given that the term "chemtrails" derives from a U.S. Air Force chemistry manual, it would be instructive to trace how this 17th-century term became attached to the present fibrous condition that has been mysteriously relegated to the psychiatrist's bible, the *DSM-V* (*Diagnostic and Statistical Manual of Mental Disorders, Fifth Edition*), as "delusional parasitosis":

> Delusional parasitosis is a rare disorder in which affected individuals have the fixed, false belief (delusion) that they are infected by "bugs": parasites, worms, bacteria, mites, or other living organisms. As with all delusions, this belief cannot be corrected by reasoning, persuasion, or logical argument. Many affected individuals are quite functional; for the minority, delusions of parasitic infection may interfere with usual activities.[4]

If I were looking for a way to cover all the bases of a covert biological experiment, I too might slip the name of a 17th-century disease into the Internet stream, then make sure it was categorized as not really there, at all.

The connection between Morgellons fibers and their chemtrails delivery system is deeply buried beneath a complicity of silence surrounding illegal international open field experiments. If medication against H1N1 or mutated swine flu virus can be sprayed by light aircraft and helicopters over 40,000

1 Mike Adams, "Strange fibers found embedded inside Chicken McNuggets." *Natural News,* August 16, 2013.

2 Megan Ogilvie, "Transgenic Animals: Genetically engineered meal close to your table." *GlobalResearch,* 22 October 2008.

3 C.E. Kellett, M.D., M.R.C.P., "Sir Thomas Browne and the Disease Called the Morgellons." *Annals of Medical History,* n.s., VII (1935), 467–479.

4 Kathryn N. Suh, MD, and Jay S. Keystone, MD. "Delusional Parasitosis: Epidemiology, clinical presentation, assessment and diagnosis." UpToDate.com: ". . .clinicians around the world have relied on UpToDate as their primary resource for medical knowledge at the point of care."

Ukrainians,[5] what's to prevent spraying the disease itself under similar benign auspices?

Could acute respiratory infections (ARI) like the "outbreaks" of coronaviruses SARS-CoV (severe acute respiratory syndrome, China, 2003), H1N1 (swine flu, worldwide, 2009), MERS-CoV or NCoV (Saudi Arabia, 2012), and H7N9 virus (bird flu, China, 2013) be distributed by chemtrails? The World Health Organization (WHO) reports that in 2002 alone, 3.9 million people died of ARI-related deaths, and that it was a leading cause of death in children under five.[6] Was the release of a self-propagating *Beauveria bassiana* fungus from the genetic engineering lab at Lincoln University in New Zealand intentional[7]?

In 1988, Leonard Cole—now Director of the Program on Terror Medicine and Security at the University of Medicine and Dentistry of New Jersey Center for BioDefense—wrote *Clouds of Secrecy: The Army's Germ Warfare Tests Over Populated Areas,* beginning his introduction with, "During the 1970s, Americans learned that for decades they had been serving as experimental animals for agencies of their government." Back then, the Cold War was blamed for "forcing" the U.S. military to experiment on the very people it had sworn to protect. The Soviet threat, however, was just another cover story for the pursuit of aggressive, well-funded weapons R&D (research and development). Once the Cold War story had been milked for all it was worth, the Cold War ended and a new enemy was trotted out: the Muslim *jihad* and war on terror.

The most frightening terrorist of all is proving to be our own government. As Cold War-era cartoonist Walt Kelly's character Pogo said, "We have met the enemy and he is us."

> Alexander M. Capron, who served as [Carter's] executive director of the President's Commission on Bioethics, said that under existing rules the army could be spraying over heavily populated areas, and the public would not know. [Interview, June 2, 1982] Capron's agency, the only federal commission concerned with ethical problems involving research on humans, was dissolved in 1983 [under the Reagan-Bush-Cheney *troika*].[8]

Again and again, Cole points out the Army's employment of "plausible deniability"—*lies,* in ordinary language: that the target population's health was not monitored (difficult to believe meticulous records were not kept); that the agents sprayed had nothing to do with an increase in disease; that the

5 Referenced in Chapter 3: "CIA Operated Aerial Spraying Plane Carrying 'Mutated' Virus Shot Down in China." *Pakalert,* November 30, 2009.

6 www.who.int/vaccine_research/diseases/ari/en/

7 "Genetic jailbreak: GMO fungus escapes in New Zealand." Institute for Responsible Technology, 8 May 2013.

8 Leonard A. Cole, *Clouds of Secrecy: The Army's Germ Warfare Tests Over Populated Areas.* Rowland & Littlefield, 1988.

military was "unaware" of scientific literature indicating that microorganisms might be dangerous to segments of the population (children, the elderly, etc.). Mycoplasma (bacteria) can induce pneumonia, chronic fatigue, respiratory and lupus-like flus and AIDS-like symptoms—in other words, a full-blown immune system assault.

In 1951, the Army sprayed the tracer *Serratia marcescens* over San Francisco and caused a small epidemic. Of the ten infected patients at the Stanford University Hospital, one died. The Army secretly convened a meeting to agree upon a plausibly deniable story and came up with the outbreak being "coincidental," then continued spraying *Serratia* and other agents over U.S. population centers. *Serratia* has been implicated in meningitis, wound infections, and arthritis, and infections have traveled through dialysis, blood transfusions, catheterization, and lumbar punctures. Over the next two decades, the infections increased dramatically in the San Francisco Bay area as military lies and cover-ups continued.

The *Clouds of Secrecy* chapter "The Army's Germ Warfare Simulants: How Dangerous Are They?" specifically addresses open-air testing. Biological, chemical, and radioactive agents known to be toxic and virulent have been sprayed everywhere in the United States "from coast to coast, over cities, in buildings, on roads, and in tunnels" since *Hemophilus pertussis* (whooping cough) was sprayed along Florida's Gulf Coast in 1955, and gases and hallucinogens were sprayed in Maryland and Utah in the 1960s. Zinc cadmium sulfide, a dry fluorescent powder used as a tracer in atmospheric studies, was sprayed for two decades over Iowa, Nebraska, South Dakota, and Virginia, despite the fact that military scientists knew that cadmium is toxic and accumulates in tissues. Instead, they and the hospital physicians working for them concentrated on keeping meticulous notes of symptoms, procedures, etc. It was a well-organized, multilevel operation, and the fate of the nonconsensual subjects was moot.

> The participants in the [1960 international symposium in Britain on Inhaled Particles and Vapours, followed by another in 1965[9]], like others engaged in biological warfare experimentation, did not appear invidious or sadistic. They assumed there was no danger because they wanted to believe there was none. Any suggestive evidence to the contrary was ignored, as were the few scientists who expressed skepticism. The vast majority conformed with the accepted belief system and behaved as most people do. It has always been easier to comply and acquiesce than to object and stand out.[10]

9 "Inhaled Particles VIII: History of the Symposia," *Ann. occup. Hyg.*, Vol. 41, Supplement 1, pp. xvii–xix, 1997. annhyg.oxfordjournals.org/content/41/inhaled_particles_VIII/xvii.full.pdf

10 Cole, *Clouds of Secrecy*, 1988.

The Hibakusha

The chemtrails-HAARP pump-and-dump is Manhattan Project II, making a comparison between today's Project Cloverleaf victims and the Hibakusha of World War II apt, indeed. The Hibakusha were the atomic bomb victims of Hiroshima and Nagasaki from whom medical treatment was withheld because they were under observation as "research subjects." Five points expose the parallels between their fate as nonconsensual "experimental" victims and our own chemtrails fate:

(1) *After the American genocidal assaults in 1945, the Atomic Bomb Casualty Commission (ABCC) was established in defeated Japan to observe and collect data on the Hibakusha. Subordinate to U.S. Occupation Forces, the ABCC was ordered to take advantage of the "golden opportunity" to record A-Bomb medical effects over time.* Today's Department of Defense (DoD) and DARPA, under the auspices of "national security," ride roughshod over all agencies involved in geoengineering "experiments."

(2) *U.S. Occupation Forces established the Japanese National Institute of Health (JNIH) and utilized scientists who had worked in the notorious Unit 731 of Ishii Shiro, the Japanese military's biological and chemical warfare division (BW/CW) and their collaborator the Institute of Infectious Diseases (IID).* The Ishii Shiro and IID sound a lot like the U.S. Biological Warfare Laboratories at Fort Detrick, Maryland (1943–1969) now dispersed among several dubious progeny, *including the National Cancer Institute*: the U.S. Army Medical Research and Materiel Command (USAMRMC), U.S. Army Medical Research Institute of Infectious Diseases (USAMRIID), National Cancer Institute-Frederick (NCI-Frederick), National Interagency Confederation for Biological Research (NICBR), and National Interagency Biodefense Campus (NIBC).

Similar to Nazi Paperclip[11] scientists, the U.S. military amnestied Unit 731 scientists, despite their atrocities. It is believed that the U.S. utilized BW/CW research of Unit 731 scientists during the Korean War (1950–1953). In 1953, American Korean War POWs testified to the International Scientific Commission for the Facts Concerning Bacterial Warfare in China and Korea, set up by the World Peace Council, that the U.S. had sprayed experimental biological weapons over Korea. To divert public attention from this atrocity, rumors of Korean brainwashing were spread throughout American media—rumors then used to justify MK-ULTRA and its 149 subprojects (1953–1973).[12]

(3) *From 1945 to the end of the Occupation in 1952, U.S. Forces banned publication of reports regarding the A-Bomb genocide.* The operations now being conducted

---

11 See Footnote 6 in Chapter 1.

12 Wikipedia: "Project MK-ULTRA is the code name of a U.S. government human research operation experimenting in the behavioral engineering of humans through the CIA's Scientific Intelligence Division." See John D. Marks, *The Search for the "Manchurian Candidate": The CIA and Mind Control: The Secret History of the Behavioral Sciences*. W.W. Norton & Company, 1991.

under a geoengineering umbrella have been discredited and marginalized as a "conspiracy theory."

(4) *Doctors were prohibited from communicating and exchanging clinical experience and research on the Hibakusha.* Similarly, physicians examining symptoms of those who have been "chembombed" or who report Morgellons fiber skin eruptions are at the very least uninformed and at the very most warned off from ongoing "national security research."

(5) *The Japanese government did nothing to help the Hibakusha,* just as U.S. and NATO governments do nothing to help their people. In the U.S., complicit agencies include NASA, NOAA, NIH, CDC, DOE, DOA, EPA, and DHS.

The 66 years between the Hibakusha and the HAARP-inspired Tōhoku earthquake and tsunami that have now spawned yet other generations of Hibakusha have been filled with quiet aerial assaults and "golden opportunities," bringing to mind a comment by Valtinsblog at *Invictus*:

> It is not an exaggeration to say that much of our modern history, including even the recent turn (or re-turn) of the U.S. clandestine agencies and military to the "dark side" use of torture and drugs on prisoners, had its origins in this diabolical deal made with war criminals from Japan and Nazi Germany. . .[13]

Are NATO countries conducting studies of who can and can't evolve in an ionized atmosphere loaded with nanoparticulates of heavy metals and polymers piggybacking biological experiments while consuming GMOs loaded with the same?

MKNAOMI

Like MKUltra, the ultra-secret Cold War MKNAOMI biological warfare project run by the CIA and Special Operations Division (SOD) of the U.S. Army at Fort Detrick, Maryland was supposedly terminated in 1970 but not outed until 1977 at the MKUltra hearings before the Senate Intelligence Committee and Subcommittee on Health and Scientific Research of the Committee on Human Resources. In their 2010 article "National Security Secrecy: Morgellons Victims Across the US and Europe," investigative reporters Hank P. Albarelli, Jr. and Zoe Martell seem almost to suggest that Fort Detrick's deep involvement with cancer may have entailed an open field experiment not unlike the chemtrails experiment now.[14]

In the early 1960s, MKNAOMI underwent a critical shift bearing directly upon our present crisis. According to microbiologist Dr. Hanley Watson,

---

13 "The Atomic Victims as Human Guinea Pigs," valtinsblog.blogspot.com, December 2, 2012. Recommended readings on the Hibakusha: Susan Lindee, *Suffering Made Real: American Science and the Survivors of Hiroshima* (1994); Sheldon H. Harris, *Factories of Death: Japanese Biological Warfare, 1932–45, and the American Cover-Up* (1995).

14 Hank P. Albarelli Jr. and Zoe Martell, "National Security Secrecy: Morgellons Victims Across the US and Europe," VoltaireNet.org, June 12, 2010. www.voltairenet.org/article165822.html. Also see Footnote 93.

MKNAOMI would no longer concentrate on developing paralyzing agents but instead would create a "designer disease that could render targeted groups or populations incapacitated, as opposed to immobilizing people":

> On July 1, 1969, a high-ranking Pentagon biological warfare official, Dr. Donald MacArthur, appeared before the Defense Department Appropriations Subcommittee of the U.S. House of Representatives. Dr. MacArthur told the assembled elected officials that . . .within the next 5 to 10 years it would probably be possible to make *a new infective microorganism, which could differ, in certain important aspects, from any known disease-causing organisms.* Most important of these is that it might be refractory to the immunological and therapeutic processes upon which we depend to maintain our relative freedom from infectious disease . . .[H]e informed the subcommittee that a research program to explore the feasibility of developing such a disease, *"a synthetic biological agent, an agent that does not naturally exist and for which no natural immunity could be acquired,"* would take only about 5 years to complete, and would cost $10 million. [Emphases added.][15]

Albarelli and Martell stress that "the Army's most secret experiments with laboratory-manipulated diseases were conducted in those same states where Morgellons is reported to be most prevalent: Texas, Florida, and California."[16]

The molecular biology arrangement between the CIA and the U.S. Army necessitated covert, nonconsensual dissemination of incapacitating and lethal materials. It also "required surveillance, testing, upgrading, and evaluation of materials and items in order to assure absence of defects and complete predictability of results to be expected under operational conditions," not to mention assisting "in developing, testing, and maintaining biological agents and delivery systems."[17] In other words, the operation was top-down and highly organized—*not* simply a faction gone rogue.

MKNAOMI continues to this day, perhaps even under Project Cloverleaf (or whatever the aerosols operations have been renamed). Victims are warned to keep quiet or marginalized; physicians, hospitals, and laboratories are informed that "national security" is at stake and they are to "cooperate" or else; media and Internet shills are directed to discredit or ignore Morgellons sufferers. Open-air testing of "huge clouds of bacteria and chemical particles" is still going on under deafening media silence and a heartless dearth of precautions "to protect the health and welfare of the millions of people exposed."

One thing has changed since the Hibakusha and MKNAOMI: the arrival of nanotechnology.

---

15 Ibid.

16 Ibid.

17 Memorandum from Chief, TSD [Technical Services Division]/Biological Branch to Chief, TSD "MKNAOMI" Funding. "Objectives and Accomplishments," 10/18/67; in "What was the CIA's MKNAOMI?" *Invictus*, November 14, 2010.

CARNICOM'S QUEST

With the help of the microscope these cinder-coloured animals may be made out, having two horns, round eyes, a tail which is long, forked, with the extremities, which are bent up, covered with hair. These worms are terrible to look at.

— *Journal de Médecine*, 1791

Clifford E. Carnicom is a conservative scientist willing to follow the evidence through innumerable experiments in order to confirm the direction the truth demands. His *vitae* (see Chapter 2) include having worked for the very Department of Defense whose covert technology he has toiled for two decades to decode. His first paper on Morgellons appeared in August 2006, though his experiments establishing the link to airborne samples had already been confirmed over and over again since the late 1990s.

Carnicom has been consistently stonewalled, denied, slandered, and discounted by U.S. government agencies, despite the 126+ military-industrial complex visitors to his website who confirm that the cold shoulder has not been because his work is inconsequential.[18] Representative Tom Udall, Senators Barbara Boxer and Dianne Feinstein, former President William Jefferson Clinton, former Secretary of Defense William Cohen, former Attorney General Janet Reno, former Secretary of the EPA Carol M. Browner, and former Secretary of the FAA Jane Garvey have all ignored his requests for confirmation of lab work.

Carnicom's methods of approach to airborne and body fibers included chemical testing, column chromatography, electrolysis, ninhydrin analysis, visible light and infrared spectroscopy analysis of peak absorbance wavelengths, and production of culture extracts.[19] For a decade and a half, due to a lack of funds, he could not afford to purchase the scanning electron microscope he needed to confirm that the fibers dropping from the sky and popping out of people's bodies were actually nano-biotechnology.[20] Denial of funding, refusal to replicate experiments, control over labs—many tricks of the trade are employed to lock freelance scientists out of inner science circles privy to "national security" projects. In *HAARP: The Ultimate Weapon of the Conspiracy*, Jerry Smith explains how the game is played, here citing how scientific proof of a link between electromagnetic exposure and cancers can be marginalized:

---

18 Carnicom, "Visitors to www.carnicom.com." Also, see "Official Responses to Aerosol Operation Inquiries" (no date), "O'Connell Opposes Spraying" (September 22, 2000), "EPA Perpetually 'Unaware'" (January 9, 2001), "USAF To Taylor: All Is 'Ordinary'" (February 1, 2001), "United States EPA Region 4 Also 'Unaware" (February 1, 2001).

19 Carnicom, "Morgellons: The Breaking of Bonds and the Reduction of Iron," November 3, 2012.

20 Carnicom, "Environmental Filament Project: An Introduction," July 9, 2013.

> With good reason, scientists trust a result only after it has been independently replicated. Studies that had not been repeated were ignored by the NAS-NRC [National Academy of Sciences - National Research Council] EMF panel. So without funding for replication, [Dr. Eugene] Sobel's and [Dr. Anthony] Miller's work can be ignored.[21]

Other "professional" players are in on the game, as well, such as the medical professionals who burden Morgellons sufferers with diagnoses such as *delusional parasitosis*, despite the obvious visible physical effects and ease with which samples could be obtained for analysis. Support groups and nongovernment organizations (NGOs) may have a "gatekeeper" agenda; Carnicom recommends monitoring information disseminated at conferences as well as contacts made there, such as he did after attending the 1st Annual Morgellons Disease Medical Research Conference in Austin, Texas on March 29, 2008.[22]

Carnicom has reported death threats, hacking, impersonation, fraud, libel, and threats against his livelihood to the FBI, and has had to terminate Internet message boards. In 2001, he became aware of a phone tap, then of frequent visits to his website by Internet Protocol Router Network (NIPRNet, NIPR.mil). The Defense Information Systems Agency (DISA) has a number of NIPRNet gateways protected and controlled by firewalls and other technologies ("NIPR Activity Increases," February 28, 2001). On March 31, 2001, NIPR.mil visited Carnicom's site for 10.5 hours to inspect 53 web pages ("NIPR.mil 10 1/2 Hour Visit," April 1, 2001).[23]

Seven Operations

Before diving into the research—the reader is welcome to go to the Carnicom Institute website and follow along (www.carnicominstitute.org)—let's contextualize the biological operation of Morgellons in the seven operations that, over time, Carnicom has concluded geoengineering is up to. In the 2005 documentary *Aerosol Crimes*, he listed five operations:

(1) *Weather/environmental modification.* The physical atmosphere has been changed.
(2) *Electromagnetic operations.* Ionization of the upper atmosphere is drawing charged particles into our lower atmosphere; barium particulates are being spread for more conductivity.
(3) *Military operations.* Weapons.
(4) *Biological operations.* Biotechnical delivery of biowarfare components via nanotech.
(5) *Planetary/geophysical operations.* Altered plasma means dramatic Earth changes via the troposphere and magnetosphere.

---

21 Jerry Smith, *HAARP: The Ultimate Weapon of the Conspiracy*. Adventures Unlimited Press, 1998.
22 Carnicom, "'Morgellons' — 2nd Session," April 11, 2008.
23 Renewed DISA interest in Carnicom's ELF experiments, late 2002-early 2003.

In the 2011 documentary *Cloud Cover,* Carnicom added a sixth operation: *intelligence,* namely sophisticated surveillance of everything and everyone on Earth for the sake of C4 [command, control, communications, computers].[24] A little later, he added a seventh operation: *detection of exotic propulsion systems,* namely detection of what are commonly discredited as UFOs. These operations extend to and beyond:

- Fiber optics-based C4, plus optical switching systems
- Space-based scalar SDI system that replaces the old ground-based missile defense
- Weather control in a plasma atmosphere
- Virtual warfare
- Biological/chemical warfare
- Chemical/electrical control of human behavior
- Electrical power transfer
- Plasma processing to break down nuclear waste (the components of which are being disposed of in the upper atmosphere)

The term *Morgellons* originally referenced an anomalous skin condition, but Carnicom has extended it to include *blood-borne vectors.*

> Erythrocyte (red blood cell) degradation and variation appears to occur in proportion to the severity of the condition. Furthermore, various erythrocyte modifications detected indicate that stem cell research should be incorporated within the investigation of the condition.[25]

From 1996 to 1999, Carnicom conducted a microscopic air particle count study in New Mexico and determined that the count in 1999 was significantly higher than counts in the preceding three years, due to increased spray activity overhead. Along with fibers, he collected and analyzed cloud progression photographs, telephotos of aircraft with what appeared to be spray apparati on the wings, and meteorological studies in arid environments that defy natural cloud formation, intuiting all the while that the fibers below were somehow connected with the aerosols being smeared over the New Mexico skies above. The 1998–99 influenza season—the first days of the military's full-on aerosol push—was categorized as epidemic:

> Pneumonia and influenza mortality exceeded the threshold for 12 consecutive weeks beginning January 24 through April 17, 1999, and peaked at 8.8% during the week ending March 13, 1999.[26]

---

24 This sixth operation would include *synthetic telepathy* technologies, what is commonly known as remote mind control, a topic so large as to merit another book.

25 Clifford E. Carnicom, "Morgellons Statement," May 9, 2009.

26 T. Lynnette Bremmer, *et al.* "Surveillance for Influenza — United States, 1997–98, 1998–99 and

In his November 2–4, 1999 report, "Ground Samples: Microscopic Fibers Revealed," Carnicom documents the grey chemtrails in the winter of 1998, the white chemtrails in November 1999 over eastern Oregon and Sacramento, California, and individual fibers measuring *less than one micron,*[27] with synthetic polymer fibers seemingly acting as carrier mechanisms. *But for what?*

The fibers exhibited extreme adhesiveness and elasticity with a tendency to form "kinked" wave-like forms that dissipated over time. After handling the fibers, people became ill; Carnicom advised caution.[28] The fibers dropping over Sedona, Arizona on July 10, 1999 had a *petrochemical* odor; in Oklahoma, they were chiffon-like; in Sacramento in February 2000, a white powder and granular clumps 200 microns wide fell.

By May 7, 2000, Carnicom had identified the first biological component: freeze-dried or desiccated red blood cells readily visible after being subjected to immersion oil. The discovery of red blood cells moved his studies in the direction of the U.S. military's sordid history of conducting biological and chemical warfare on its own citizens. As a precaution, he declared:

> The source material for the images presented herein has been duplicated and distributed to numerous locations across the United States, and it is secured by various methods. The ramifications of this recent discovery establish sufficient cause for widespread involvement of the American people in this issue, and for subsequent criminal investigations and Congressional hearings.[29]

Having established a relationship between biologicals and the aerosols dropping them, Carnicom turned his inquiry toward nailing down the relationship between the airborne sub-micron fibers and the fibrous Morgellons structures emerging from bodies. Albarelli and Martell describe the symptoms of Morgellons sufferers as "the discomforting sensation of insects crawling on and biting or stinging their skin":

> This sensation results in skin lesions that can appear much like mild to severe cases of acne. The lesions can appear anywhere in a patient's body and quite often contain fiber-like strands or fibrous material. *The fibers are the most perplexing visible feature of Morgellons.* Often when an attempt is made to remove or extract the fibers the material will resist and act to withdraw or move away from whatever instrument is being employed.[30] [Emphasis added.]

1999–00 Seasons," Centers for Disease Control and Prevention, *Morbidity and Mortality Weekly Report*, October 25, 2002.

27 Compare with acrylic fiber = 12 microns wide, human hair = 5–6 microns, nylon = 12–15 microns, polyester = 14 microns, silk = 15 microns, spider web = 7 microns, wool = 15–25 microns, cotton = 10 microns.

28 See Maryna van Wyk, "Strange, sticky, wiry threads similar to spider's web falls in the Karoo." *Rapport*, South Africa, Cape Edition, 25 June 2000.

29 Carnicom, "Biological Components Identified," May 11, 2000.

30 Albarelli and Martell, "National Security Secrecy: Morgellons Victims Across the US and Europe," 2010.

Innumerable sufferers' observation that the fibers "resist and act to withdraw or move away"—like parasites—would prove crucial.

The Morgellons condition is neither neurotic excoriation nor delusional parasitosis nor a "matchbox sign."[31] It is real, entails real pain and suffering, and is connected to the chemtrails-HAARP pump-and-dump. In fact, *Morgellons is bioengineered and is being delivered by geoengineering.* Carnicom's characterization of Morgellons, born of years of research, is worth quoting at length:

> The term "Morgellons" refers to a condition that was originally perceived to manifest primarily as an anomalous skin condition. The visible symptoms commonly include skin lesions that resist healing and the presence of unusual filaments that emanate from sores and the skin in general. . .
>
> More recent research strongly indicates the underlying symptoms are much deeper and more broadly distributed than has been realized, and that blood-borne vectors may be a common denominator amongst affected individuals . . .Erythrocyte (red blood cell) degradation and variation appears to occur in proportion to the severity of the condition . . .
>
> The presence of skin anomalies as the primary criterion for determining the existence of the condition appears to be especially deficient, and it is recommended that blood-borne conditions amongst the general population be investigated in addition to any skin manifestation in the minority of the population. The existence of the condition is now acknowledged by the Centers for Disease Control, the National Institutes of Health, and the Mayo Clinic.[32]

THE FIVE "DIMENSIONS" OF GENETICALLY ALTERED MORGELLONS PATHOGENS

By late 2007, the form, size, and structure of unusual airborne filaments had been confirmed as being linked to Morgellons skin and dental fibers as well as to anomalies in human blood samples, one being from an individual manifesting advanced Morgellons symptoms.[33] Manifestations in the blood and skin of victims exhibiting Morgellons symptoms are identical, and these are identical to the condition of blood cells in samples *of those who don't manifest Morgellons,* causing Carnicom to comment:

31 A dismissive psychiatric term that refers to the meticulous collections of fibers that sufferers pluck from their skin.

32 Carnicom, "Morgellons Statement," May 9, 2009.

33 See high-magnification images at Carnicom, "Morgellons: Airborne, Skin & Blood — A Match," December 10, 2007; and "The Biggest Crime of All Time," March 1, 2011.

> The question of Morgellons manifestation may be one of degree . . .It has been stated that the Morgellons condition may have a much broader basis and distribution than we might like to admit or know.[34]

In other words, in Morgellons victims the filaments are *in extremis*, but the truth is that *all* populations subject to the chemtrails delivery system will probably experience at least a slow degradation of health as the immune system is undermined. Carnicom's consistent challenge has been to find living subjects with *normal* hemoglobin—meaning that the "organism" may be in everyone's blood. Pathogens in the blood and blood cells is abnormal, the blood being a sterile environment; nevertheless, it appears that the blood of his samples is undergoing a transformation: the cellular structure is changing to a more *fibrous* form, and spherical structures like those inside the fibers are appearing in disturbed blood cells.

As to the fiber's nature, function, and purpose, Carnicom first tested for bacteria, fungi, viruses, parasites, prions, *Rickettsiae*, and *Chlamydiae*. His analysis had to meet the following criteria: (1) Sub-micron (0.5–0.7 microns); (2) intracellular; (3) associated with respiration, given that chronic respiratory ailments point the way; (4) spherical to oblate but pleomorphic (as in mycoplasmas) as well; (5) Gram stain produces gram-negative results; (6) present in diverse physical samples; and (7) *Chlamydiae* genus illnesses should correlate with Morgellons-like symptoms.[35] Size and structure immediately eliminated the virus, and size alone the parasite (eukaryotic forms are 10–100 microns).

In 2010, the Carnicom Institute acquired both visible light and infrared spectrophotometers, opening up a broader range of spectral analyses—the fingerprints or unique signatures within a particular range of frequencies of chemical substances or species. The spectral analysis of the airborne environmental filament that Carnicom had sent the EPA a decade before (only to have it returned unidentified) revealed a distinctive signature perfectly matching the absorbance spectral analysis of a one-year-old culture developed from an oral filament sample of a Morgellons victim. One culture after another, human and airborne, the matches were exact, including *DNA*: "The same degree of similarity has been achieved with a culture developed from a human DNA extraction."[36]

All evidence has pointed to four major forms and one minor form of the genetically altered Morgellons pathogen being delivered by aerosols, genetically modified foods, water systems, and even piggybacked on vaccinations and other inoculations. After years of examining and re-examining samples of pathogens from skin, hair, scalp, teeth and gums, saliva, urine, ears, and blood, Carnicom admits that humanity is facing an imminent health holocaust:

---

34 Carnicom, "Morgellons: Airborne, Skin & Blood — A Match," December 10, 2007.
35 Carnicom, "And Now Our Children," January 11, 2008.
36 Carnicom, "Morgellons: pH, Conductivity, Ions & Live Analysis," January 10, 2010.

> The vitality and viability of human existence and life on this planet, as it has been known to exist, is under threat.[37]

Now, three years later, he has gone so far as to say:

> This work demonstrates that the "Morgellons" *situation* has been completely understated and underestimated in its significance and distribution. It is no longer to be considered as unique to any life form or species. The term itself, as commonly interpreted to represent a condition or disease, is inadequate to encompass the scope of impact to the biology of the planet.[38]

The characteristics *or dimensions* are listed here in the order of observation, not necessarily importance:

(1) *The bounding filament* delivery system, 12–20 microns diameter; exhibits as luminescent when airborne, in skin lesions and dental samples, but perhaps not in the blood. It houses the fibrous submicron network (2); however, in one gum sample, Carnicom discovered that it held the *Chlamydia*-like organisms he now calls *cross-domain bacteria* (3), indicating the possibility that the various forms *morph*.

With chemicals and a spectroscope, Carnicom ascertained that the external casing of environmental filament samples is composed of keratin, an especially impervious protein structure rather like a hair. With sodium hydroxide and gentle heat, he was finally able to penetrate a sample fiber from Serbia: the original fiber material was pure white, whereas the two internal colored effects were yellowish and reddish.[39] Appearance to the contrary, the encasing filament is not a fungus.

(2) *Submicron interior fibrous network* inside the bounding filament (much like fiber-optics cable), 0.7 microns in diameter; fungus-like, morphologically similar to hyphae (fungal filaments), leading to fungal overgrowths like Candida. Resists all chemical and heat extremes.

(3) *Cross-domain bacteria,*[40] originally categorized as *spherical or oblate structures*, 0.5–0.7 microns. Though mycoplasma-like and *Chlamydia*-like, especially *Chlamydia pneumoniae*, it is neither. Intracellular form primarily in blood samples; the degree of blood cell damage corresponds to the number of cross-domain bacteria in contact with or adjacent to other blood cells. *These forms are the precursors of all the other forms.* Resists all chemical and heat extremes.

(4) *Hybrid* form, pleomorphic, ribbon- or sausage-like.[41] Mycoplasma-like, it is seemingly a tertiary stage of development that resists all chemical and heat extremes.

---

37 Carnicom, "The Breath of A Decade," December 18, 2010, edited January 10, 2011.

38 Carnicom, "The New Biology," January 18, 2014.

39 Carnicom, "Environmental Filament: Keratin Encasement," January 7, 2013; "Environmental Filament Penetration," January 6, 2013.

40 Carnicom, "Cross-Domain Bacteria Isolation," January 18, 2014.

41 Carnicom, "Morgellons: A Status Report," October 8, 2009.

(5) *"Budding" structures* on the edge of the fiber at irregular intervals indicative of a *growth* or *reproductive process* and may be related to the spread of the disease.[42]

CULTURES

Carnicom has successfully and repeatedly cultured the Morgellons pathogen. The significance of being able to culture a pathogen means it might be possible to control, inhibit, reduce, or eliminate similar pathogenic forms in the body.[43]

> . . .the culture forms represent a viable means of study of metabolism and biochemical structure that holds numerous advantages over attempting to study these same processes within the human body.[44]

From examination of cultures holding four of the structures, Carnicom determined that:

1. The culture flourishes in an acidic environment that the pH of a red wine medium does not affect;
2. Conductivity increases with the growth of the culture, indicating that ion concentration increases with growth;
3. Filament growth passes through three phases: first white, then green, then black;
4. Full growth cycle is two to three months;
5. Chloride ion concentration increases, perhaps because the cross-domain bacteria "eats" ferrous ion, which in turn bears upon the degradation of the red blood cell integrity noted in blood samples.

By constantly adjusting culture mediums—agar, red and white wines, simulated wine, broths—Carnicom has been able to study pH, conductivity, ion analyses, nutrients, inhibitors, etc. For example, with a time-lapse video under the microscope (450X), Carnicom watched a cultured growth of 50 microns an hour on top of a dental sample placed in a bouillon agar medium, with the growth in length accompanied by increased density and complexity.[45] Adding iron sulfate enhances growth, and adding hydrogen peroxide to white wine and iron sulfate dramatically increases growth in hours instead of days with wine alone.

---

42 Photographs at 700–8600X of all four types and the "budding" structures are available at "Morgellons: Pathogens & The General Population," April 9, 2008.

43 Carnicom, "Culture Breakthrough," July 12, 2008; "Culture Work Is Confirmed," August 18, 2008.

44 Carnicom, "Morgellons: Infrared Spectroscopy — Culture Confirmation," January 1, 2013.

45 Carnicom, "Morgellons: Growth Captured," August 21, 2008. Detailed photographs of the culture process can be found in "Morgellons: A Status Report," October 8, 2009.

The hydrogen peroxide points to Fenton's reaction[46] and the formation of the hydroxyl (free) radical, one of the most reactive oxidants known—oxidation being the process by which atoms, molecules, or ions lose electrons. Why this might be important to the Morgellons challenge is indicated here:

> In cells and tissues, such particles can attack a host of surrounding biomolecules to produce new free radicals, which, in turn, attack yet other compounds. Thus, the formation of a single free radical can initiate a large number of chemical reactions that are ultimately able to disrupt the normal operations of cells.[47]

Given that the hydroxyl radical is an expected product of our metabolism, conditions for expanded growth of the cross-domain bacteria in its various stages is likely, which means serious health issues. Therefore, the identification of specific chemical and biological conditions underlying Morgellons symptoms is of the highest priority.

Growth flourishes in acidity and with oxidizers like sodium hypochlorite or bleach, sodium chlorite or MMS, calcium hypochlorite or MMSII. *Alkalinity, on the other hand, inhibits growth.* Changing the pH of the growth environment might not kill the cross-domain bacteria, but it will make it go dormant. Its presence alone seems to increase acidity, and given that iron corrodes quickly in an acidic environment, the blood cells have a harder and harder time absorbing oxygen—all in addition to the structural damage of the blood. (The pH of physical death is always acidic.)

When Carnicom subjected oral filament cultures in the usual wine medium to *blue light*, the growth rate was explosive, going from a filmy layer to dense filaments of greater diameter *in just 24 hours*. Thus the application of frequencies of signature harmonics (the speed of an electromagnetic wave) plays a significant role in the cross-domain bacteria's growth and health of the host. More on frequencies later in this chapter.

Proteins and Amino Acids

By means of column chromatography, Carnicom ascertained the existence of a *protein* in culture growths (beyond the bounding filament keratin sheath), which pointed to DNA. Proteins, the building blocks of nature, are now *patentable* by Big Pharma.[48] Discerning two related complexes—the *iron-protein complex* and the *iron-dipeptide complex*—he considered three alternatives:

(1) the similarity to a dimorphic fungal-like organism;

---

46 In 1894, H.J.H. Fenton discovered that certain metals have a strong catalytic power to generate highly reactive hydroxyl radicals (.OH). The iron-catalyzed hydrogen peroxide is thus Fenton's reaction.

47 Theodore L. Brown, *Chemistry: The Central Science*. Pearson-Prentice Hall, 2006.

48 For extensive proteins analysis, see Carnicom, "Morgellons Research: Proteinaceous Complex Identified," March 14, 2012.

(2) a joint existence of bacterial-like and fungal-like organisms in symbiotic relationship; or

(3) the specter of a genetically created or designed organism.[49]

Dipeptides are the combination of two amino acids and constitute a primitive form of protein development. In late 2012, Carnicom confirmed the presence of the amino acids cysteine and histidine and posited that deficiencies and disturbances in just these two amino acids might be behind the high oxidation levels and joint pains that Morgellons sufferers complain of.[50] With the reducing agents ascorbic acid (Vitamin C), N-acetyl cysteine (NAC), and glutathione, Carnicom was able to interfere with the molecular bonding of the iron-dipeptide complex and reduce oxidation:

> It will be found that there are important interactions and relationships in the body between cysteine compounds, NAC and glutathione. The combination of influences and interactions between iron, cysteine, histidine, ascorbic acid, N-acetyl cysteine and glutathione represents an important pathway of research for the Morgellons condition.[51]

The Tree of Life

In 2009, something defying "all conventional understanding of blood cell development" occurred. When Carnicom broke down the external casing of a dental/gum filament with chemistry and heat (strong alkalis of sodium hydroxide and bleaches, plus hydrochloric acid and boiling), inside he found *artificial or deliberately modified blood cells.* He recognized that erythrocytes were artificial because they were just too perfectly formed, and their hostility to reconstructive chemicals and heat seemed fiercely programmed. Advanced technologies in stem cells and genetic transfers were at work. But inside the artificial erythrocytes were the telltale sub-micron structures associated with Morgellons.[52]

Growing erythrocytes in a test tube was once "considered to be the holy grail of biological achievement with huge implications for bioengineering, human health and the human species."[53] Researchers have been able to sustain and perpetuate existing blood cells in a growth medium, but not create *new* cells. This is surely the cutting edge of stem cell research.

Carnicom employed three erythrocytic detection methods for "developing modified erythrocytes (red blood cells) within cultured [Morgellons] dental samples": (1) direct microscope observation (8000–10000X); (2) the

49 Carnicom, "Morgellons and Recent Findings," January 2012.
50 Carnicom, "Amino Acids Verified," November 3, 2012.
51 Carnicom, "Morgellons: The Breaking of Bonds and the Reduction of Iron," November 3, 2012.
52 Carnicom, "Artificial Blood (?)," August 27, 2009.
53 Ibid.

Kastle-Meyer presumptive test commonly used for forensic blood identification; and (3) the HEMASTIX (TMP) presumptive forensic test.[54] It was this discovery that made Carnicom realize that *blood should be the focal point of Morgellons study, not skin*, given that erythrocyte degradation is buried at the core of this biological assault on humanity.

In 1977, based upon their genetic relationships (the RNA world hypothesis), microbiologist and physicist Carl R. Woese whittled down the time-tested six biological kingdoms (Eubacteria, Archaebacteria, Protista, Fungi, Plantae, Animalia) to three *domains*: Bacteria, Archaea, and Eukarya, all of which vary according to cell type, cell wall, membrane, protein synthesis, transfer of RNA, and sensitivity to antibiotics.

It is from Woese's domains that Carnicom derived the name "cross-domain bacteria" to describe Morgellons structures as being somewhere between plant and animal: *Morgellons structures like the cross-domain bacteria do in fact cross the lines of all three domains.*[55] Like Archaea (found in volcanic vents, under ice shelves, etc.), chemical, heat, and cold (-50° to -60°C) stresses impact neither the vitality, growth, nor reproduction of interior structures, all of which share cell metabolism, size, pathogenic impact, and symptomatology with Bacteria; and the fungus-like bounding filament is of the Eukarya domain. The erythrocytic "new blood" forms, however, challenge all of the above boundaries.

On February 5, 2010—two days after Carnicom posted this classification paper on his website, DARPA publicly disclosed that its BioDesign project ($6 million)—plus $20 million to the synthetic biology program, and $7.5 million to sequencing, analyzing, and functionally editing cellular genomes—was seeking "to re-write the laws of evolution":

> . . .living, breathing creatures that are genetically engineered to "produce the intended biological effect" . . .fortified with molecules that bolster cell resistance to death, so that the lab-monsters can "ultimately be programmed to live indefinitely."[56]

Genetically programmed locks on "tamper-proof" cells, a kill switch—and not a peep in mainstream media about DARPA's insistence upon taking a God-like direction. As one DARPA employee told *Wired* Danger Room: "I would love to comment, but unfortunately DARPA has installed a kill switch in me." Not a joke, I fear, given all the microbiologists who have had bizarre "accidents."[57]

---

54 Carnicom, "Blood Issues Intensify," April 22, 2009.

55 Carnicom, "Morgellons: A New Classification," February 3, 2010.

56 Katie Drummond, "Pentagon Looks to Breed Immortal 'Synthetic Organisms,' Molecular Kill-Switch Included." *Wired*, February 5, 2010.

57 See "Dead Scientists And Microbiologists — Master List," compiled by Mark J. Harper. Rense.com, February 5, 2005.

The Cross-Domain Bacteria Eats Iron

Sub-micron Archaea-like cross-domain bacteria—smallest of all the growth forms—feed on iron. (In hostile environments, many Archaea feed on iron and sulfur.) They enter the serum of the blood and breach the outer wall of erythrocytes (red blood cells). Metabolic imbalance follows.[58]

Much of Big Oil's R&D goes into perfecting iron-eating bacteria.[59]

In 2011, Carnicom did extensive work on the Morgellons pathogen's affinity for the iron in human blood. The following lengthy quote from his October 15, 2011 Abstract thoroughly apprises the reader of the frightening possibility that the Morgellons pathogen may have been created and programmed to devour iron in order to compromise the iron-oxygen basis of biological life:

> A substantial body of research has accumulated to make the case that the underlying organism (i.e. pathogen) of the so-called "Morgellons" condition, as identified by this researcher, is using the iron from human blood for its own growth and existence. It will also be shown that the bio-chemical state of the blood is being altered in the process. The implications of this thesis are severe as this alteration affects, amongst other things, the ability and capacity of the blood to bind to oxygen. Respiration is the source of energy for the body.
>
> This change is also anticipated to increase the number of free radicals and to increase acidity in the body. This process also requires and consumes energy from the body to take place; this energy supports the growth and proliferation of the organism. The changes in the blood are anticipated to increase its combination with respiratory inhibitors and toxins. The changes under evaluation may occur without any obvious outward symptoms. It is also anticipated that there are consequences upon metabolism and health that extend beyond the functions of the blood. This change represents essentially a systemic attack upon the body, and the difficulties of extinction of the organism are apparent. Physiological conditions that are in probable conjunction with the condition are identified. Strategies that may be beneficial in mitigating the severity of the condition are enumerated.
>
> In summary, I now see five major challenges before us with the "Morgellons" issue, based upon the research that I have conducted to date:
>
> 1. The iron within the blood, to a partial degree, is being changed in a way that it no longer binds with oxygen at the normal levels that are expected. The organism uses iron to sustain its existence and growth. Diminished oxygen-carrying capacity of the blood is therefore expected in coincidence with the severity of the condition.

58 Carnicom, "Morgellons: In the Laboratory," May 22, 2011. This paper provides an excellent review of Carnicom's research up to that point.

59 Ben Li, "Iron eating bacteria deciphered" (*The Gauntlet*, June 3, 2004); Pippa Wysong, "Metal-Eating Bacteria Corrode Pipes in Oil Industry" (*Access Excellence*, September 25, 2004).

2. The presence of free radicals is likely to increase in number and extent as a result of the oxidation process. Free radicals are known to wreak havoc in the living system.
3. The altered iron (Fe3+ versus Fe2+) now binds to other molecules than oxygen, many of them toxic or harmful to health. Several of these alternative ligands are known respiratory inhibitors, and therefore further exacerbate the failures in respiration.
4. In addition to the consumption of iron already identified, the bacteria-like form, which appears to be at the origin of the pathogen, binds to oxygen to support its own existence. This combination further increases the severity of consequence to human health.
5. The presence of the organism, as encountered, appears to be extensive throughout the body, at a minimum occurring in the circulatory, digestive, and urinary systems.[60]

The late 2012 acquisition of a Perkin Elmer 1320 infrared spectrophotometer gave him a window into the molecular structure of organic compounds and confirmed yet again that:

- cultures from oral samples are identical to oral filament samples[61];
- penetration of the Serbian environmental filament revealed internal structures of a pure white original filament, one yellowish, and one strong red, pointing to an erythrocytic red;
- the external casing is composed of keratin, an especially impervious protein structure.[62]

Chemtrails carry polymers as well as conductive metals and crystalline substances. We are breathing and ingesting polymers and polyethylene-silicon-carbon nanofibers and nanowires that can house and/or piggyback combinations of pathogens, blood cells, sedatives, and nanoparticulates programmed to be microprocessors and sensors—all of which pass into our blood and bypass the blood-brain barrier.[63] These Morgellons self-assembling, self-replicating proteins replete with sensors, antennae, wires, and arrays are in our food, water, and bodies where they copy the DNA of pathogens, cancer cells, etc., and increase them. They are able to create pseudo-hair, pseudo-skin, and chimeric forms that look like insects and parasites.

60 Carnicom, "Morgellons: A Thesis," October 15, 2011; edited December 1, 2011, and May 10, 2013.
61 Carnicom, "Morgellons: Infrared Spectroscopy — Culture Confirmation," January 1, 2013.
62 Carnicom, "Environmental Filament Penetration," January 6, 2013.
63 Jim Giles, "Nanoparticles in the Brain." Nature.com, January 5, 2004.

TO YOUR IMMUNE SYSTEM!

If, however, millions of people are already on prescription pharmaceuticals to "calm them down" [long-term, what is this doing to their ability to think clearly?] and, in addition, are breathing poisoned air rife with mind-distorting chemicals, then how clearly (if at all) is anyone able to think? How can anyone feel well and safe if the air we breathe is deliberately poisoned and is affecting our ability to think cogently? It is like Diogenes, the ancient Greek, searching for a truthful individual. No one seems to have the desire, or courage, or authority to stop this massive poisoning, because it is the secret plan of the elite insiders to deliberately destroy everything we once knew.

— Dr. Ilya Sandra Perlingieri, "Chemtrails: The Consequence of Toxic Metals and Chemical Aerosols on Human Health," *Global Research,* May 12, 2010

That the chemicalized, processed food of the convenient American lifestyle spells death and medical/Big Pharma industry profits is no news flash by now. Only the poorest and most downtrodden, ad-addicted adults continue to vote with their shrinking dollar to eat death.

But the situation described in this book demands a serious review of what we in the West have been conditioned to think of as "lifestyle." Many lifestyle decisions are actually cultural assumptions and habits, all of which must now be reexamined in the light of the global challenge facing the embattled immune system. Chemtrails, the ionized atmosphere synergizing new pollutants, and the increasing ubiquity of GMO foods laced with biotech sensors inside and outside our bodies may, in fact, be the straw that breaks our lifestyle. Barring a worldwide insurrection, the choices left us veer between self-victimization replete with constant complaining and enriching the medical/Big Pharma industry, and learning how to wield the biological laws that govern an acid/alkaline balance to strengthen our assaulted immune systems.

Growing healthy, whole food will be the number one radical act of the 21st century. Under our ionized aluminum skies, whole forests are dying and soil becoming so alkaline that plants are prematurely turning yellow. Insiders like Monsanto may have gotten the jump on us with aluminum-resistant seed and patents, but savvy farmers and gardeners are awakening and exploring covered area options to protect plants.

Besides *greenhouses,* another option is rising up. Abandoned urban warehouses are being turned into greenhouses. Farmed Here LLC, the largest *vertical farm* in the United States, is growing stacked "boutique greens" (herbs, lettuce, microgreens, and edible plants like beets and sunflowers for their sprouts) under a warehouse roof and over an "aquaponic" farm (water circulating or misting under the plants) of tilapia and other fish whose excrement fertilizes the plants. Whereas normal greenhouses can still depend upon diffuse sunlight, vertical farms are exploring energy options like LEDs, solar, wind, and methane to run the artificial lighting, heat, and water that such urban farming needs.[64]

In Asia, Indian rice and potato farmers are harvesting world-record yields with the SRI (system of rice intensification) technique: 22.4 tons of rice on 2.5 acres of land without GMO seed or chemicals. Economist and Nobel laureate Joseph Stieglitz visited the fields and declared that the farmers were "better than scientists":

> For many Westerners . . .it's difficult to separate the concept of "progress" from its inevitable modifier, "technological." SRI may not be technology-based, but it's science-based and sophisticated. It's also continually field-tested and improved through farmers' own feedback. It's exactly the kind of flexible, responsive system you'd demand from any truly sustainable agriculture—as opposed to the regimented, top-down application of chemical- and biotech-based approaches.[65]

Carnicom's six strategies for countering the Morgellons organism assaulting our red blood cells and devouring the iron in our blood are:

1. Alkalinize the blood
2. Anti-oxidation
3. Increase the utilization and absorption of existing iron
4. Inhibit the growth of iron-consuming bacterial-archeal-like forms
5. Improve the flow of bile to further alkalinize the blood and aid the digestive process.
6. Detoxify the liver.[66]

None of these strategies subscribe to the take-a-pill school of health. Every strategy points to the necessity of daily preparation of nutritious whole foods to build healthy iron-rich blood and strengthen the organs that help the blood to rebuild. Certainly, a vitamin C supplement (or fruit) will help to increase iron absorption and produce hemoglobin, plus provide an antioxidant, but a daily dose of vitamin C needs the backup of a healthy eating regimen.

In the late 1960s, I began studying the macrobiotic way of life with Michio Kushi and Herman Aihara. For almost half a century, I have followed the

---

64 Martha Irvine, "In a Chicago suburb, an indoor farm goes 'mega.'" *Seattle P-I*, March 28, 2013. Also, microbiologist Dickson Despommier's *The Vertical Farm: Feeding the World in the 21st Century*.

65 Tom Laskawy, "Miracle grow: Indian farmers smash crop yield records without GMOs," grist.org, 22 February 2013.

66 Carnicom, "Morgellons: In the Laboratory," May 22, 2011.

yin/yang (acid/alkaline) approach to eating to encourage my intestinal tract and organs to maintain a strong immune system in an increasingly polluted environment. The beauty of the acid/alkaline yin/yang law is that it can free us from the medical/Big Pharma establishment by teaching us to *alter our blood chemistry with how and what we eat.* My last visit to a physician was when I was 13; I am now 66 and still committed to health as a lifestyle choice, discipline, and deep pleasure.

The idea that freedom might mean living in accord with ancient biological laws is foreign to youthful Western medicine, which clings to drugs and surgery. Many Western women define freedom as being free of traditional roles like cooking, and many men look down on food preparation as no more transformative than filling up at a gas station. With the quest for money and career taking precedence over quality food preparation, Westerners graze and forage at restaurants and fast-food outlets whose bottom line is profit, not health. Even "health nuts" spurn biological laws, insisting that their body tells them what it needs. Meanwhile, excess (fat) and internal disorders soar in tandem with medical/Big Pharma profits.

A liver-gall bladder cleanse[67] will do the immediate job of flushing the liver and purging gallstones to increase the flow of bile to help alkalinize the blood and aid digestion, but at the same time it will wipe out colonies of helpful gut (intestinal) flora[68], which even with careful eating will take weeks to rebuild. As Aesop's fable "The Hare and the Tortoise" teaches, *slow and steady wins the race*: Better to slowly build and maintain a healthy gut system than just depend upon flushing it out now and then.

Speaking of detox, 3–10 grams (dry) of daily unrefined seaweeds—especially kelps like kombu, wakame, macrocystis, nereocystis, focus, and ascophyllum—armor the immune system against thermal and nonthermal radiation. If you are concerned about the impact that Fukushima radioactive isotope releases might have on Pacific-grown seaweed, contact the Seaweed Stewardship alliance (hand harvesters in California, Oregon, and Washington) for updates. Kombu has 1,000–8,000ppm of measurable iodine.[69]

Daily brown rice well chewed massages the intestinal tract and keeps it clean while alkalinizing the blood and providing nutrients. Organic vegetables, cooked or raw, also provide stimulating roughage (cellulose) and nutrients. If your digestive tract is weak, lightly sautéed, steamed, or baked veggies will be easier to digest than raw and well worth the loss of a few vitamins. For protein, organic beans are excellent. However, they must be pre-soaked, cooked thoroughly

67 Many sites like curezone.com/cleanse/liver/ offer recipes and advice.

68 Wikipedia: "The human body carries about 100 trillion microorganisms in its intestines, a number ten times greater than the total number of human cells in the body. The metabolic activities performed by these bacteria resemble those of an organ, leading some to liken gut bacteria to a 'forgotten' organ."

69 Ryan Drum, Island Herbs, "Radiation Protection Using Seaweeds." Southwest Conference on Botanical Medicine, April 13–14, 2013.

before adding salt, then chewed well, given that they are a protein-carbohydrate combination.

As for eating animals and their dairy products—nonorganic and organic—ask yourself, What does it mean for my immune system that fish, birds, and animals are embattled, as well? If you do choose to eat animals and animal products, eat little and low on the food chain (chicken or fish). Humans are the only creatures that eat dairy after being weaned. While many insist on dairy as a quick grazing-foraging source of protein, calcium, etc., dairy is hard on lungs already hard-pressed to breathe heavy metals and polymers loaded with nanosensors.

An over-acid condition and weakened immune system go hand in hand. The health benefits of alkalinizing are well supported in health literature regarding the benefits of antioxidants (ascorbate, glycerin, ester salts or sodium citrate, and garlic compounds like ally/cysteine, alliin, allicin, and allyl disulfide).[70] Sugar, vinegar, and wine produce acidity, as do all refined flour products and "junk food." Avoid acid-producing foods or be prepared to immediately counterbalance them. This may even include naturopathic medications or boosters like the following:

- *Chlorella*, a single-celled algae, expels cadmium and prevents it from poisoning the liver
- *Astragalus root (huang qi)* opens the lungs; the dry cough that follows forces the cilia to expel particulates
- *GABA (gamma-aminobutyric acid*) clears the lungs
- *Nigella sativa (blackseed)* strengthens the immune system
- *Artemisium (Artemisinin or qinghaosu*[71]), the active compound in wormwood, works to kill the malaria parasite, cancer cells, and the Morgellons organism, all of which sequester iron in the blood
- *Glutathione*, an amino acid that assists in detoxifying; available as a nebulizer, intravenous (by prescription), and sublingual powder
- *Zeolite powder and food-grade diatomaceous clay* for clearing heavy metals
- *Milk thistle (silibinin*[72]) kills skin cells mutated by UVA radiation and protects against damage by UVB radiation; recommended for diabetes, Hepatitis C, and Morgellons
- *Baking soda baths* alkalinize the skin
- *Negative ion generator*

---

70 Carnicom, "Morgellons: A Discovery and A Proposal," February 22, 2010, edited June 12, 2011; "Morgellons: Growth Inhibition Confirmed," March 15, 2010.

71 Naomi Ishisaka, "UW Scientist Henry Lai Makes Waves in the Cell Phone Industry." *Seattle Magazine*, January 2011; Lai and Singh, "Magnetic-field-induced DNA strand breaks in brain cells of the rat." *Environmental Health Perspectives*, May 2004; 112(6): 687–694.

72 "Silibinin, Found in Milk Thistle, Protects Against UV-Induced Skin Cancer." *ScienceDaily*, January 30, 2013.

Copious information and tips can be found at chemtrails and Morgellons sites (see Resources) as well as at LymeBuster Chat.[73] Share what works, keeping in mind that each body and diet is different. As you browse and experiment, continue fine-tuning your baseline regimen and chew well, given that chewing alkalinizes what slides down into the alchemical, miracle-working intestinal tract.

Even the wireless environment we are all so attached to *acidifies the blood.* Wireless Internet emits radiation, whether on or off, including cell and cordless phones or any other wireless transceiving devices and their towers or bases. Nonionizing microwave radiation may not produce much heat but it is ten million times as strong as the average natural background and is known to eventually produce cancer and a raft of maladies pointing to a compromised immune system. An Italian study found that herpes viruses like the Epstein-Barr virus (EBV) genome in latently infected human lymphoid cells are reactivated via daily exposures to ambient electromagnetic radiation, *including house wiring* (50Hz in Europe, 60Hz in the U.S.). Headache, muscle twitching, and skipped heartbeats can arise from a pulsing 60Hz electric field, and DNA can be modulated by a foreign magnetic field, meaning any frequency not our natural Schumann resonance.[74] As Dr. Robert O. Becker explained:

> In most parts of the US the local utility company delivers plenty of health-damaging high frequencies that ride into your home on your 60-cycle household current. RF currents also enter your home via water and gas pipes, phone lines, etc. Once inside, they jump from surface to surface, even to things like wooden furniture, as they like to spread out over surfaces. Unless you check with an electric field meter, you will not be aware of their presence and they may cause the above symptoms. I have no doubt that at the present time the greatest polluting element in the earth's environment is the proliferation of electromagnetic fields.[75]

Therefore, include environment in the tally of your immune system lifestyle. For example, home WiFi diminishes with distance, but routers designed to handle multiple computers at neighborhood schools do not. Children spend 180 days per year near routers radiating 5.8GHz so that each router can communicate with multiple computers, plus 2.4GHz for the 10–18 WiFi

---

73 Lyme is named after Lyme, Connecticut, where the illness was identified in 1975. A bacterium transmitted through infected deer ticks, Lyme shares symptoms with Morgellons, and Ginger Savely, DNP, of Austin, Texas (www.gingersavely.com) believes that Lyme sufferers have weaker immune systems and may therefore be more vulnerable to Morgellons.

74 Grimaldi, S. *et al.* "Exposure to a 50 Hz electromagnetic field induces activation of the Epstein-Barr virus genome in latently infected human lymphoid cells." *Journal of Environmental Pathological Toxicological Oncology*. 1997; 16(2-3):205–7.

75 Robert O. Becker, *Cross Currents: The Perils of Electropollution, The Promise of Electromedicine*, 1990.

routers. Given that spinal cords and DNA act as antennae,[76] children with their soft bones and brainpans are especially vulnerable.

Another hidden factor is the synergy between ELFs and some pharmaceuticals:

> Synergisms between pharmacological agents and endogenous neurotransmitters are familiar and frequent. The present review describes the experimental evidence for interactions between neuropharmacological compounds and the classes of weak magnetic fields that might be encountered in our daily environments. Whereas drugs mediate their effects through specific spatial (molecular) structures, magnetic fields mediate their effects through specific temporal patterns. Very weak (microT range) physiologically patterned magnetic fields synergistically interact with drugs to strongly potentiate effects that have classically involved opiate, cholinergic, dopaminergic, serotonergic, and nitric oxide pathways. The combinations of the appropriately patterned magnetic fields and specific drugs can evoke changes that are several times larger than those evoked by the drugs alone. These novel synergisms provide a challenge for a future within an electromagnetic, technological world. They may also reveal fundamental, common physical mechanisms by which magnetic fields and chemical reactions affect the organism from the level of fundamental particles to the entire living system.[77]

Finally, radio wave sickness—electrohypersensitivity (EHS) or electrosensitivity (ES)—is often the very biological stress that produces the acidic environment in which the Morgellons pathogen thrives. EHS, "a bona fide environmentally inducible neurological syndrome,"[78] was publicly acknowledged to be a physical impairment in Sweden in 2000, then the UK followed suit in 2005:

> Special cables are installed in sufferers' homes while electric cookers are replaced with gas stoves. Walls, roofs, floors and windows can be covered with a thin aluminum foil to keep out the electromagnetic field—the area of energy that occurs round any electrically conductive item.[79]

---

76 "Investigation of the spinal cord as a natural receptor antenna for incident electromagnetic waves and possible impact on the central nervous system," *Electromagnetic Biological Medicine*, Vol. 31, No. 2, 101–111, June 2012; "DNA is a fractal antenna in electromagnetic fields," *International Journal of Radiation Biology*, Vol. 87, No. 4, 409–415, April 2011.

77 P.D. Whissell and M.A. Persinger, "Emerging Synergisms Between Drugs and Physiologically Patterned Weak Magnetic Fields: Implications for Neuropharmacology and the Human Population in the Twenty-First Century." *Current Neuropharmacology* December 2007; 5(4): 278–288. Persinger has spent years experimenting with weak EM fields on brains. Wikipedia: "During the 1980s [Persinger] stimulated people's temporal lobes artificially with a weak magnetic field to see if he could induce a religious state (see God helmet)."

78 Andrew Marino *et al.*, "Electromagnetic hypersensitivity: evidence for a novel neurological syndrome." www.ncbi.nlm.gov/pubmed/21793784?dopt=Abstract

79 Sarah-Kate Templeton, "Electrical Fields Can Make You Sick." *The Times OnLine*, 11 September 2005.

Many American EHS refugees are fleeing the 322 million cell phone subscribers and millions of wireless laptops, tablets, and modems and heading for havens like Green Bank, West Virginia, where wireless is outlawed, thanks to the Radio Quiet Zone needed by the Green Bank Telescope.[80]

Protective products are slowly trickling into the market. EMF protective diodes and bio-field protectors are now available. (See Resources.) French fashion's Smuggler label, in collaboration with the XLIM Institute in Limoges, France, has developed a suit that blocks EM waves by interweaving nonallergic nickel, stainless steel, aluminum, and faux gold into jacket pockets,[81] and EM-SEC Technologies (www.emsectechnologies.com) is producing a roll-on or spray-on Wireless Security Coating to block radio signals.[82]

In conclusion, we have begun to see that Morgellons is not a few thousand people with skin lesions but bioengineered fibers being scattered far and wide whose inhalation and ingestion threaten a major breakdown of the entire body system:

> On a macro scale, we can see that some of the more obvious issues to be addressed concern iron disruption, amino acid presence and protein rebuilding, acidity, oxidative stress, availability of oxygen, thyroid and metabolism issues, halogen toxicity and substitution concerns, joint and skeletal integrity and elasticity, blood and cellular integrity, and potential neural disruption. Unfortunately, the list is not exhaustive but it is representative of some of the health concerns that have been brought to the forefront and reported on.[83]

Until more scientists and health professionals are on board and doing all that needs to be done regarding Morgellons not just under the skin but in the blood, keep Hering's Law of Cure in mind:

> *Symptoms of a chronic disease disappear in a definite order, going in reverse and taking about one month for every year the symptoms have been present.*
> *Symptoms move from the more vital organs to the less vital organs; from the interior of the body toward the skin.* (Every organ has a frequency.)
> *Symptoms move from the top of the body downward.*

Poor comfort, indeed, but this drama has just begun to unfold. Hear Carnicom's clarion call and get your immune system sea legs under you.

80 "Wireless Refugees: 'Cell Phones, WiFi Making Us Sick.'" Wusa9.com, February 27, 2013.

81 "French fashion firm develops suit that blocks electromagnetic waves." Agence France-Presse, March 21, 2013.

82 But not for scalar ground wave signals. For those, one needs separated layers of metal sheeting similar to Reichian technology. Doors and windows may require the same.

83 Carnicom, "Morgellons: A Working Hypothesis: Neural, Thyroid, Liver, Oxygen, Protein and Iron Disruption," August 12, 2013.

VISITORS TO WWW.CARNICOM.COM

Let it be noted that United States government computer systems are to be used for official purposes only.

1. Desert Research Institute in Nevada (weather modification research institution) (repeat visits)

2. Fort Lewis Army Military Base (now Joint Base Lewis-McChord), Washington state (home of Special Forces air squadron)

3. Lockheed Martin (U.S. aviation and space defense contractor) (repeat visits) (repeat repeat visits)

4. Los Alamos National Laboratory (LANL) (repeat visit)

5. Allergan Pharmaceutical Corporation (allergy pharmaceutical research)

6. Alliant Techsystems (Space and Strategic Defense Systems contractor)

7. Raytheon Defense Systems (U.S. defense contractor) (repeat visit) (repeat repeat visit) (repeat repeat repeat visit)

8. Boeing Aircraft Company (100 visits minimum)

9. United States Defense Logistics Agency (supplies and support to combat troops)

10. Davis-Monthan Air Force Base, Tucson AZ (home of 355th Wing) (repeat visits) (repeat repeat visits) (repeat repeat repeat visits)

11. Dept of Defense Naval Computer and Telecommunications Area Master Station

12. U.S. Naval Sea Systems Command

13. Western Pacific Region of the Federal Aviation Administration, Lawndale CA (repeat visit) (repeat visit) (repeat visit)

14. National Aeronautics and Space Administration (NASA) Langley Research Center (ten visits minimum)

15. United States Environmental Protection Agency (EPA) (20 visits minimum)

16. St. Vincent Hospital, Santa Fe, New Mexico

17. Headquarters United States Air Force, The Pentagon

18. United States Department of the Treasury (repeat visit) (repeat visit)

19. United States Department of Defense Educational Activity

20. Andrews Air Force Base, Proud Home Of Air Force One

21. United States Federal Aviation Administration (FAA)

22. United States Naval Research Center, Washington D.C.

23. Rockwell-Collins (U.S. defense contractor)

24. Honeywell (U.S. defense contractor) (repeat visit)

25. Wright-Patterson Air Force Base, Dayton OH (repeat visit) (repeat repeat visit)

26. Kadena Air Force Base, Okinawa, Japan

27. Camp Pendleton, United States Marine Corps (mandatory U.S. Defense anthrax vaccination program described at www.cpp.usmc.mil) (repeat visit) (repeat visit)

28. Ames Research Center, NASA (a primary mission is to research ASTROBIOLOGY, i.e. the study of life in outer space) (repeat visit)

29. Space Dynamics Laboratory, Utah State University, North Logan, Utah

30. Merck (Pharmaceutical Products and Health Research) (repeat visit)

31. McClellan Air Force Base, Sacramento, CA. (The Sacramento Air Logistics Center at McClellan Air Force Base, California performs depot maintenance on the KC-135 Stratotanker aircraft and is heavily involved in space and communications-electronics.) (repeat visit)

32. TRW (U.S. defense contractor) (repeat visit)

33. Teledyne Brown Engineering (U.S. defense contractor)

34. United States Navy Medical Department

35. Air National Guard, Salt Lake City, Utah

36. Monsanto Company (chemicals, pesticides, and pharmaceutical products) (repeat visit) (repeat repeat visits)

37. U.S. Department of Veterans Affairs

38. ARCO Chemical Company (now under LyondellBasell Industries)

39. Sundstrand Aerospace (U.S. defense contractor)

40. National Oceanic and Atmospherics Administration Aeronomy Laboratory (conducts fundamental research on the chemical and physical processes of the Earth's atmosphere)

41. Allied Signal Corporation (chemical, aerospace, energy) (repeat visit) (repeat repeat visit) (repeat repeat repeat visit) (repeat repeat repeat repeat visit)

42. Aviation Weather Center, National Oceanic and Atmospherics Administration (NOAA)

43. United States Army Medical Department (repeat visit)

44. NASA Goddard Space Flight Center

45. Applied Physics Laboratory, a research division of Johns Hopkins University, which supports the U.S. Defense Department

46. United States Naval Health Research Center, San Diego, CA

47. Headquarters, United States Army, The Pentagon

48. United States General Accounting Office (the investigative arm of Congress; GAO performs audits and evaluations of government programs and activities.)

49. Bristol-Myers Squibb Company (Pharmaceutical Research and Development)

50. United States Naval Criminal Investigative Service (a worldwide organization responsible for conducting criminal investigations and counterintelligence for the Department of the Navy and for managing naval security programs)

51. National Computer Security Center (NCSE) (involved in advanced warfare simulation)

52. The Mayo Clinic (repeat visit) (repeat repeat visit) (repeat repeat repeat visit)

53. The Federal Judiciary (home of the United States Supreme Court)

54. United States Federal Emergency Management Agency (FEMA) (controls a comprehensive, risk-based, emergency management program of mitigation, preparedness, response and recovery) (repeat visit)

55. United States Naval Surface Warfare Center, Crane IN (repeat visit) (repeat repeat visit)

56. United States National Guard Public Affairs Web Access (no public access to this site)

57. United States Senate (repeat visit) (repeat repeat visit) (repeat repeat repeat visit) (repeat repeat repeat repeat visit)

58. Headquarters, United States Air Force Reserve Command

59. Kaiser Permanente (integrated managed care consortium), Oakland, CA

60. United States Naval Warfare Assessment Station

61. Air University, United States Air Force

62. United States Naval Research Laboratory (repeat visit)

63. Enterprise Products Partners LP (MTBE production)

64. United States Navy Naval Air Weapons Stations, China Lake CA

65. California Pacific Medical Center

66. United States Defense Information Systems Agency (Mission: "To plan, engineer,

develop, test, manage programs, acquire, implement, operate, and maintain information systems for C4I and mission support under all conditions of peace and war.")

67. Harvard Pilgrim Health Care (runs FDA's Sentinel, a program "to track the safety of drugs, biologics and medical devices")

68. San Francisco Department of Public Health

69. BJC Health System, St. Louis, Missouri

70. United States Open Source Information Systems (OSIS; an unclassified confederation of systems serving the intelligence community with open source intelligence) OSIS sites include:

(AIA) Air Intelligence Agency, Kelly AFB, San Antonio, TX

IC-ROSE (CIA) Central Intelligence Agency, Reston, VA

(DIA) Defense Intelligence Agency, Washington, D.C.

(NSA) National Security Agency, Ft. Meade, Laurel, MD

(NIMA) National Imagery & Mapping Agency, Fairfax, VA

(NAIC) National Air Intelligence Center, Wright-Patterson AFB, Dayton, OH

(NGIC) National Ground Intelligence Center, Charlottesville, VA

(MCIC) Marine Corps Intelligence Center, Quantico, VA

(NMIC) National Maritime Intelligence Center, Office of Naval Intelligence, Suitland, MD

(ISMC) Intelink Service Management Center, Ft. Meade, Laurel, MD (repeat visit)

71. New Mexico Department of Health

72. United States Space and Naval Warfare Systems Command (SPAWAR)

73. United States McMurdo Research Station, Antarctica

74. Orlando Regional Healthcare System, Florida

75. United States Andersen Air Force Base, Guam

76. United States Misawa Air Base, Japan

77. United States Hickam Air Force Base, Hawaii

78. United States Osan Air Force Base, Korea

79. Royal Air Force, Lakenheath, Suffolk, UK

80. United States Scott Air Force Base, Illinois

81. United States F.E. Warren Air Force Base, Wyoming

82. United States Air Force News Agency

83. United States Langley Air Force Base, Virginia (repeat visit)

84. United States Tinker Air Force Base, Oklahoma

85. United States McConnell Air Force Base, Kansas

86. United States Charleston Air Force Base, South Carolina

87. United States Randolph Air Force Base, Texas

88. United States Air Force Reserve Command, Robins Air Force Base, Georgia

89. United States Seymour Johnson Air Force Base, North Carolina

90. United States Bolling Air Force Base, Washington, DC

91. United States Keesler Air Force Base, Mississippi

92. United States Hill Air Force Base, Utah

93. United States Vandenberg Air Force Base, California

94. United States Minot Air Force Base, North Dakota

95. United States Eielson Air Force Base, Alaska

96. Andrews Air Force Base, Proud Home Of Air Force One (repeat visit)

97. Headquarters United States Air Force, The Pentagon (repeat visit) (Visitors 75–96 arrived within a 24-hour period on 09/23/99)

98. United States Cannon Air Force Base, New Mexico

99. United States McGuire Air Force Base, New Jersey

100. United States Beale Air Force Base (home of the U-2 fleet of reconnaissance aircraft)

101. United States Department of Justice Federal Bureau of Prisons

102. Metnet — United States Navy (weather reporting system and SPAWAR)

103. TRADOC — United States Army Training and Doctrine Command, Fort Monroe, VA

104. *Newsweek* magazine

105. United States Defense Advanced Research Projects Agency (DARPA)

106. Massachusetts Medical Society, owner/publisher of *New England Journal of Medicine*

107. Office Of The Secretary Of Defense: The Office Of William S. Cohen, Secretary Of Defense (repeat visit)

108. Headquarters United States Air Force, The Pentagon (repeat repeat visit)

109. United States Joint Forces Command (reports to U.S. Secretary of Defense) (repeat visit)

110. Naval Warfare Assessment Station, Corona CA

111. Los Angeles County Emergency Operations Center

112. Commander in Chief, United States Pacific Fleet, United States Navy

113. Headquarters United States Air Force, The Pentagon

114. Defense Logistics Agency, Administrative Support Center in Europe

115. United Stated Department of Defense Network Information Center, Vienna, VA (repeat visits)

116. Office of the Assistant Secretary of the Army

117. Headquarters, United States Air Force, The Pentagon (repeat visit)

118. *U.S. News and World Report*

119. Naval Air Warfare Center — Aircraft Division (repeat visits)

120. New Zealand Parliament

121. Headquarters United States Air Force, The Pentagon (multiple repeat visits)

122. NIPR — Department of Defense Network Operations (NIPRNet); the Defense Information Systems Agency (DISA) has established a number of NIPRNet gateways to the Internet protected and controlled by firewalls and other technologies. (repeat visits)

123. Peterson Air Force Base, Colorado Springs, CO (home of NORAD and SPACECOM)

124. Raytheon (visits immediately after introduction of HAARP implications)

125. United States Army War College

126. Lawrence Berkeley National Laboratory

127. Fermi National Accelerator Laboratory

CONCLUSION

Look Up!

Under the HAARP program the USA is creating new integral geophysical weapons that may influence the near-Earth medium with high-frequency radio waves . . .[T]he significance of this qualitative leap could be compared to the transition from cold steel to firearms, or from conventional weapons to nuclear weapons. . . [The HAARP program] will create weapons capable of breaking radio communication lines and equipment installed on spaceships and rockets, provoke serious accidents in electricity networks and in oil and gas pipelines, and have a negative impact on the mental health of people populating entire regions.

— International Affairs and Defense Committee, Russian State Duma, 2002

Chemtrails are creating the chemical corridors for HAARP's electromagnetic waves to follow. Laid daily over all 28 NATO nations, chemtrails linger for hours, spreading out into *cirrus contrailus* cloud cover in the name of weather engineering and amelioration of "global warming" or "climate change."

The truth, however, is more ominous. Geophysical, chemical, and biological warfare "experiments" are being quickened by a pump-and-dump action of High-frequency Active Auroral Research Project (HAARP) ionospheric heater technologies and aerosols loaded with nanoparticulates of heavy metals, polymers, and bioengineered submicrons, iron-devouring pathogens that eventually filter down into the soil and the bodies of all living creatures.

Though weather control appears to be the objective of the chemtrails-HAARP combination, the overarching agenda is to keep the atmosphere unnaturally charged so as to maintain *full spectrum dominance* and *C4*—command, control, communications, and computers—over all of Planet Earth.

"Extreme weather"—floods, droughts, snowstorms in spring, heat waves in winter, tornadoes in Colorado, 9.0 earthquakes, hurricanes, etc.—is indeed due to HAARP and other ionospheric heaters around the globe blasting the ionosphere and ionizing our atmosphere with microwaves for various political, economic, and biological agendas.

By steering the jet stream and ground-based directional arrays in a highly ionized atmosphere, weather can be made to draw rain from one area to another, or just as easily dry up moisture and produce extreme drought. Needless to say, being able to manipulate weather as a weapon of war is a considerable global power chip. Besides increasing big bank/IMF debt-servicing, Wall Street weather derivatives, and reconstruction profits at the expense of devastated communities, a drought, hurricane, tsunami, or flood can terrify small developing nations into submission or serve as an instrument of blackmail forcing nations to behave in ways they otherwise might not.

Between HAARP "pumping" the ionosphere and chemtrails "dumping" conductive metal particulates like barium, a charged conductive atmosphere is perfect for a C4 wireless world loaded with ready and waiting transmitters and receivers: GWEN and cell (microwave) towers and radio observatories, satellite platforms, power lines, buried fiber optics cables, phased arrays, NEXRADs, etc. Civilians think only of their cell phone and laptop convenience, but the military is planning for OOTW (operations other than war).

While big defense contractor telecommunications corporations and wireless drones hoover up information on everyone, chemtrails-HAARP work together to add a whole other dimension to eavesdropping: wireless nanosensors in the bodies and brains of everyone everywhere quietly erasing the lines between war and peace, soldier and civilian, inner and outer. "Owning the weather" is only one part of full spectrum dominance; the other part is a C4 that includes intimate access to each and every person and population.

The overall decline in human health points to what this military thrust is doing to our immune systems. Breathing an ionized atmosphere of aerosols delivering nanosized particulates of conductive metals and polymers loaded with sensors and microprocessors would be enough to tip biological health over the edge, but add to this the expectation that all biological life forms are to serve as Petri dishes for biotech "cross-domain bacteria" falling from the sky and protruding from Morgellons skin sores and it becomes obvious that an *assault* is underway. Every day, every night, we are breathing and ingesting what is being dropped on us.

Every era has its powerful sociopaths and military mindsets that call for creativity and will to fight to uphold the best of what it means to be human. If life itself is being loaded like a gun, it is our responsibility to think through the fight before us for the sake of future generations who prefer being human to being "enhanced" biomachines. Let's take on that responsibility by looking up and learning how to cloud-watch our *War of the Worlds* skies.

CLOUD-WATCHING

We are all now destined to become weather-watchers. Developing a clear mental picture as to how ionospheric heaters like HAARP work can help you make sense of today's skies. Picture the heater boiling the ionosphere and a high-pressure area pushing the stratosphere up along with our atmosphere, displacing the jet stream sometimes hundreds of miles from its normal flow. These high-pressure zones are always dry and create a bumper zone to block moisture-carrying lows that would normally arrive as rain. Iran recently complained that rains expected in 2012 were deliberately diverted to Europe. Low-pressure (cold) and high-pressure (hot) winds can also be increased, storms strengthened as well as steered, etc. "Global warming," "climate change," and "extreme weather" stem from this "force multiplier" technology.

Look up into the ionized skies overhead and you may see *lightning with no thunder*. Such a sight may become common in a barium- and aluminum-filled atmosphere being pumped as a nearby tornado is steered.

When attempting to discern the difference between contrails and chemtrails, keep in mind NASA's insistence that contrails may persist for a maximum of one hour, and that only the rare "persistent contrail" will spread to 2 km with a length of 60 km (37 mi). "Typically, contrails can only form at temperatures below -76°F and at humidity levels of 70 percent or more at high altitudes, according to NOAA meteorologist Thomas Schlattes. Even in the most ideal conditions, a jet contrail lasts no more than 30 minutes."[1]

If you want to track and log aircraft doing flyovers in your area, you might order a PiSkytracker; or you may have an Android or iPhone app that allows you to interface with online services like www.FlightRadar24.com, Flight Aware or Virtual Radar Server.

The "clouds" we are now watching have a density, form, and behavior far beyond the water vapor and dust of yesteryear, and far more unusual textures and patterns than can be offered by typical aircraft contrails. Watch the jets laying chemtrails back and forth in a grid and you know you are witnessing something other than the straight destination flight patterns of commercial airlines. Some even turn the aerosol off and on in a dash-dot pattern. Then there are the "cottage cheese" and "tortoise shell" patterns, cross-thatching and loops that are *not* natural alto-cumulus, just as the daily veil of cloud cover over some areas is not a natural cirro-stratus. The now almost routine circular rings around the Sun are from hexagonal crystals in the aerosol mix. Put on Polaroid glasses or sunglasses and study the colored refracted edges around the "clouds."[2]

In 2002, A.K. Johnstone (see Chapter 3) explained how weather operations generally worked:

> During a chemtrail "weather-creating" flyover, extremely large Xs (markers) are formed in the sky at high altitude, 30,000 feet (9,000 m). They are accompanied by repetitious linear trails laid out from east to west, south to north, or vice versa. Within a few minutes of chemtrail placement, cirrus clouds begin to form from

1 Bob Fitrakis and Fritz Chess, "Stormy Weather." *Columbus Alive*, December 6, 2001.
2 Clifford E. Carnicom, "Synthetic Clouds Revealed," June 22, 2000.

> the trails, gradually changing into layers of clouds (cirrostratus) and, finally, darkening into storm clouds (cirrocumulus), precursors of precipitation.[3]

The aerosol spraying is not necessarily uniform, given that weather is not the only operation going on. From season to season, certain geographic areas may be clear while others are inundated. For example, in spring and summer 2013, reports indicated that spraying was light in South Dakota, Wisconsin, and Oklahoma while spraying in Ontario and Quebec was heavy; in August, Arizona, Nevada, and Indiana experienced some relief while Missouri, Massachusetts, and upstate New York were heavily sprayed.

The heavy aerosol spraying of the Denver area in August 2013 could have been in preparation for the September 9–13 flooding of the Denver-Fort Collins area, home of North American Aerospace Defense Command and U.S. Northern Command headquarters (NORAD/USNORTHCOM) in Denver; the Air Force Academy north of Colorado Springs; Buckley Air Force Base in Aurora; Peterson Air Force Base and Schriever Air Force Base in Colorado Springs; and Fort Carson south of Colorado Springs.

Clifford Carnicom shares two key traits of chemtrail aerosols: the asymmetric "core tracks" (ribbon-like, possibly filamentous in nature) *inside the aerosol trail,* and the repeated presence of characteristic "pulse" emissions—like "donuts on a rope." The core tracks provide the "rope" for subsequent pulse "donuts" to hang on. The fact that the "rope" maintains its form for some time indicates the presence of "a discrete substance," and the fact that the "discrete substance" does not immediately fall (as real precipitation would), but *expands,* indicates the presence of an extended physical-chemical reaction unrelated to evaporation, with the reaction increasing in intensity due to ionization and perhaps desiccant qualities.

Pay attention to your symptoms and those of your animals and environment. If chemtrails are being pulsed to stimulate weather fronts, you may perceive it in your environment or body

Parents, be aware that your children are being re-educated. Not only did the U.S. Postal Service produce a "Cloudscape" stamp series to document artificial clouds,[4] but government schools are conditioning children to view chemtrails as "persistent contrails" that create *cirrus contrailus* clouds. On the "My NASA Data" page, an entire "Contrail Watching for Kids" lesson plan is ready for teachers, replete with an illustration of a chemtrail-clotted sky.[5] And "friendly sky" drones are just Robby the Robot's cousin. For only $3,999, Robots LAB-BOX's AR Drone the Quadcopter, Sphero the Robotic Ball, ArmBot the Robotic Arm, or Mobot the Mobile Robot will come to your neighborhood elementary school to teach core concepts in algebra, physics, geometry, trigonometry, and calculus.

---

3 A.K. Johnstone, *Defense Tactics: Weather Shield to Chemtrails.* Hancock House, 2002.
4 stampcenter.com/blog/2009/08/12/cloudscapes-postage-stamps/
5 mynasadata.larc.nasa.gov/804-2/contrail-watching-for-kids/

Look up and take up this fight for humanity, never doubting that commitment, will, and high-mindedness are superior to the mindset of those consumed by lust for power, those who would destroy others' humanity as they have allowed their own to be destroyed.

GLOSSARY

ACID pH <7

ACOUSTIC GRAVITY WAVES Infrasound low frequency (20–100Hz) modulated by *(groans and hums)* ultra-low infrasonic waves (0.1–15Hz)

ALBEDO Latin, "whiteness." Reflected sunlight; the diffuse reflectivity or reflecting power of a surface

ALKALINE pH >7

AM Amplitude modulation

ANTHROPOGENIC Man-made

ATMOSPHERE Composed of the *troposphere* (6–20 km), stratosphere (20–50 km), mesosphere (50–80 km), thermosphere (80–690 km)

AURORAL ELECTROJET The ionospheric curved charged particle stream at high latitudes, as opposed to the *equatorial electrojet*

BOLOMETER Measures incident electromagnetic radiation by heating a material with a temperature-dependent electrical resistance; from the Greek *bole* (βολή) for something thrown, as a ray of light

CATION An ion or group of ions having a positive charge

CIA Central Intelligence Agency

CME Coronal mass ejection

dB Decibel; a measurement used in acoustics and electronics, such as gains of amplifiers, attenuation of signals, and signal-to-noise ratios

DIA Defense Intelligence Agency

ELF Extremely or extra low frequency, 3–300 Hz

EPA Environmental Protection Agency

ERP Effective radiated power

FAA Federal Aviation Administration

FM Frequency modulation

FSB Frequency-selective bolometer; low noise sub-millimeter to mid-infrared sensor

GAUSS (G) Measurement of magnetic flux density or magnetic induction

GE/GM/GMO Genetically engineered / genetically modified organism

GIGAHERTZ GHz, one billion (1,000,000,000) cycles per second (Hz)

HERTZ Hz, cycles per second (cps)

HUMINT Human intelligence (CIA, DIA)

HYDROSCOPY Optical device used for viewing objects far below the surface of water

HYGROSCOPY A substance's ability to attract and hold water molecules from the surrounding environment. *Hygroscope*: instrument showing changes in humidity

IMINTI magery intelligence (NGA)

IONOSPHERE From the top of the mesosphere to the top of the thermosphere

INTERFEROMETRY HAARP is a scalar interferometer. See Myron W. Evans, P.K. Anastasovski, T.E. Bearden *et al.*, "On Whittaker's Representation of the Electromagnetic Entity in Vacuo, Part V: The Production of Transverse Fields and Energy by Scalar Interferometry" (*Journal of New Energy*, 4(3), Winter, 1999, 76–78); and www.padrak.com/ine/JNEV4N3.html

KILOMETER 0.621371 miles (a little over half a mile)

LIDAR Light Detection and Ranging / Laser Imaging Detection and Ranging

MAGNETOSPHERE Overlaps the ionosphere and extends into space to 60,000 km (37,280 miles) toward the Sun, and over 300,000 km (186,500 miles) away from the Sun (nightward) as the Earth's magnetotail

MASER Microwave Amplification by Stimulated Emission of Radiation

MASINT Measurement and signature intelligence (DIA)

MEGAHERTZ MHz, one million (1,000,000) cycles per second (Hz)

MICRON Micrometer; one-millionth of a meter

NANOMETER A billionth of a meter

NASA National Aeronautics and Space Administration

NGA National Geospatial-Intelligence Agency

NOAA National Oceanic and Atmospheric Administration

NONIONIZING Nonthermal radiation; insufficient energy to cause ionization *radiation* or heating: electric and magnetic fields, radio waves, microwaves, infrared, ultraviolet, and visible radiation

NRO National Reconnaissance Office (housed at one time in Room 4C1052 of the Pentagon)

ONI Office of Naval Intelligence

ORGONE Wilhelm Reich's term for the life-giving ether

PM Particulate matter; particles 10 microns or less, 1 micron being 1/70 the thickness of a human hair. Coarse particle = 10,000–2,500 nanometers; fine particles = 2,500–100 nanometers; nanoparticle (ultrafine) = 1–100 nanometers

ppm Parts per million

PLASMA Fourth state of matter; an ionized or electrically conductive gas with an abundance of charged and free electrons in it

RADAR Radio Detection And Ranging

RFR Radio frequency radiation

SCALAR Having only magnitude, not direction

SDI Strategic Defense Initiative; "Star Wars"

SIGINT Signals intelligence

SRM Solar Radiation Management

SYNERGY Increased intensity caused by the combination of two or more substances

TTA Tesla tech array

BIBLIOGRAPHY

Bearden, T.E. *The Final Secret of Free Energy* (1993)

- *Cancer and the Unresolved Health Issues in the Biological Effect: EM Fields and Radiation* (1993)

- *Toward A New Electromagnetics: Part I: The Solution to Tesla's Secrets and the Soviet Tesla Weapons* (1991)

- *Toward A New Electromagnetics: Part II: Reference Articles for Solution to Tesla's Secrets* (1991)

- *Gravitobiology: A New Biophysics* (1991)

- *Star Wars Now: The Bohm Aharonov Effect, Scalar Interferometry, and Soviet Weaponization* (1988)

- *Fer De Lance: A Briefing on Soviet Scalar Electromagnetic Weapons* (1986)

- *New Tesla Electromagnetics and the Secrets of Electrical Free Energy Proof of Free Energy Devices and Supporting Data* (1984)

Beason, Ph.D., Colonel (USAF Ret.) Doug. *The E-Bomb: How America's New Directed Energy Weapons Will Change the Way Future Wars Will Be Fought.* Da Capo Press, 2005. [*Caveat lector:* HAARP is not mentioned.]

Becker, Dr. Robert O. *Cross Currents: The Perils of Electropollution, The Promise of Electromedicine.* New York: Jeremy P. Tarcher/Penguin, 1990.

- *The Body Electric: Electromagnetism and the Foundation of Life.* William Morrow, 1998.

Begich, Dr. Nick and Jeane Manning. *Angels Don't Play This HAARP: Advances in Tesla Technology.* Earthpulse Press, 1995, 2002.

Begich, Dr. Nick and James Roderick. *Earth Rising: The Revolution.* Earthpulse Press, 2000.

- *Earth Rising II: The Betrayal of Science, Society and the Soul.* Earthpulse Press, 2003.

Begich, Dr. Nick. *Controlling the Human Mind: The Technologies of Political Control or Tools for Peak Performance.* Earthpulse Press, 2006.

Bertell, Rosalie. *Planet Earth: The Latest Weapon of War: A Critical Study into the Military and the Environment.* The Women's Press, Ltd., 2000.

"Biological Testing Involving Human Subjects By the Department of Defense, 1977." Hearings Before the Subcommittee on Health and Scientific Research of the Committee of Human Resources, 91st Congress 1st Session, March 5 and May 23, 1977.

Brown, Thomas J. *Loom of the Future: The Weather Engineering Work of Trevor James Constable.* Borderland Sciences, 1994.

Cathie, Bruce L. *The Harmonic Conquest of Space*, 1998 [free on the Internet at issuu.com/hunabkuproductions/docs/harmonic-conquest-of-space---bruce-cathie]

- *The Energy Grid: Harmonic 695, the Pulse of the Universe [The Investigation into the World Energy Grid*], 1997

- *The Bridge to Infinity: Harmonic 371244*, 1989

Cole, Leonard A. *Clouds of Secrecy: The Army's Germ Warfare Tests Over Populated Areas.* Rowman & Littlefield Publishers, 1988.

Herman, Michael. *Intelligence Power in Peace and War.* Cambridge University Press, 1996. [Former senior British Intelligence officer at GCHQ.]

Johnson, Andrew. "Re-Investigating Climate Change: Why it is NOT being caused by increased CO_2 emissions from human activity," December 27, 2010. www.checktheevidence.com.

Kirby, Peter A. *Chemtrails Exposed.* $.99 at Kindle, 2nd Ed., 2012.

MacDonald, Gordon J.F. "Geophysical Warfare: How to Wreck the Environment." *Unless Peace Comes: A Scientific Forecast of New Weapons.* Ed. Nigel Calder. A. Lane, 1968. pp 181–205.

Ponte, Lowell. *The Cooling.* Englewood Cliffs, NJ: Prentice-Hall, 1976.

Rauscher, Elizabeth and William van Bise. "Fundamental Excitatory Modes of the Earth and Earth-Ionosphere Resonant Cavity." *Harnessing the Wheelwork of Nature: Tesla's Science of Energy.* Ed. Thomas Valone. Adventures Unlimited Press, 2002. pp 233–268.

Smith, Jerry E. *HAARP: The Ultimate Weapon of the Conspiracy.* Adventures Unlimited Press, 1998.

- *Weather Warfare: The Military's Plan To Draft Mother Nature.* Adventures Unlimited Press, 2006.

Sweetman, Bill. *Aurora: The Pentagon's Secret Hypersonic Spyplane.* Motorbooks, 1993.

Thomas, William. *Chemtrails Confirmed.* Bridger House, 2004.

- *Scorched Earth: The Military's Assault on the Environment*, Chapter 11. New Society, 1994.

Weather As A Force Multiplier: Owning the Weather in 2025. August 1996.

Wood, Judy, Ph.D. *Where Did The Towers Go? Evidence of Directed Free-Energy Technology on 9/11*, 2005.

INDEX

RESOURCES

AUDIOVISUALS

- [*Aerosol Crimes*] *Cloud Cover: The Deliberate Destruction of Our Atmosphere*. Concerned Citizens for Clear Skies, 2011.

Best Evidence: Chemical Contrails. Discovery Channel. Season 1, Episode 3 aired April 6, 2007. At this writing, up on YouTube in five segments.

Chemtrails: Mystery Lines in the Sky. Dir. William Thomas and Paul Grignon. 2000. 27 minutes. www.canadaskywatch.com/articles/documentary/2012/08_09_mystery_lines_in_the_sky.html

Farmageddon. Dir. Kristin Marie, 2011. How the USDA is in cahoots with Big Agribiz.

Full Signal. Dir. Talal Jabari, 2010. Scientists around the world discuss their research into health effects related to cell towers and phones.

Genetically Modified Society: Operation Paul Revere. Dir. Julio N. Rausseo, 2013. Free on the Internet

Griffin, G. Edward, Michael J. Murphy, Peter Wittenberger. *What in the World Are They Spraying?* Truth Media Productions, 2010. Free at www.youtube.com/watch?v=jfokhstYDLA

Holes in Heaven? HAARP and Advances in Tesla Technology. Dir. Wendy Robbins, narrated by Martin Sheen. UFOTV.com, 2005. Free on the Internet.

"Increases in Extreme Weather, Food Prices, and Illness: The Unspoken Connection," a videoed presentation for pilots and flight crew by doctoral researcher David Lim of Reading University, March 2013.

"Invisible Hazards in the Wireless Age (The Chemtrails Song)" by Trillion, 2009.

Murphy, Michael J. and Barry Kolsky. *Why in the World Are They Spraying?* Truth Media Productions, 2012. Free at Farm Wars farmwars.info/?p=9095

"Morgellons" — Introductory Remarks Clifford Carnicom and Gwen Scott, ND March 21, 2008 www.carnicom.com/morgvid1.htm

"Morgellons" — 2nd Session

April 11, 2008

www.carnicom.com/morgvid2.htm

Resonance: Beings of Resonance. Dir. James Russell and John K. Webster, 2012. www.youtube.com/watch?v=5vb9Rox_oNQ

Primer in conjunction: "Resonance Phenomena in 2D on a Plane," www.youtube.com/watch?v=Qf0t4qIVWF4

Skywatcher: Contrails, Chemtrails, and Artificial Clouds, 2012. Thirty-minute documentary video posted on YouTube. www.artificialclouds.com

The Great Culling: Our Air. Dir. Paul Wittenberger and Chris Maple. Framing the World Productions, www.thegreatculling.net/

— *The Great Culling: Our Water*. (The fluoride "social experiment")

Thunderbolts of the Gods, 2006. www.thunderbolts.info/wp/

Toxic Sky? Hosted by Paul Moyer. KNBC, Los Angeles, May 23, 2006. On the situation in Los Angeles County and in the Coastal Ranges of

Northern California. Transcript of video commentary and interview with Rosalind Peterson at www.holmestead.ca/chemtrails/nbc-toxicsky.html

Toxic Skies. Dir. Andrew C. Erin. Gravitas Ventures, 2008. Hollywood film. World Health Organization physician uncovers the real cause of an epidemic is the secret government chemtrail program. Full film at www.youtube.com/watch?feature=player_embedded&v=St40If-Y_so

We Feed the World by Austrian filmmaker Erwin Wagenhofer about globalization and the food industry, 2005. Now only at trutube.tv/video/7811/We-Feed-the-World-full

WEBSITES

Chemtrails

agriculturedefensecoalition.org — Rosalind Peterson / Agriculture Defense Coalition [California Skywatch, Southern California Skywatch, the Bonnefire Coalition, and Plumas Skywatch are all in partnership with the Agriculture Defense Coalition].

aircrap.org "Monitoring the *Planned* Poisoning of Humanity"

www.checktheevidence.com

www.chemtrailsfactorfiction.com

chemtrailsplanet.net Geoengineering Exposed: Global Warming Linked to Advanced Climate Modification Technology

www.chemtrailsproject.com The Chemtrails Awareness Project.

www.commoncrime.net — Anthony J. Hilder

www.freeworldfilmworks.com/actionary.html

thecontrail.com

www.coalitionagainstgeoengineering.org — Mike Murphy

www.data4science.net — Ted Twietmeyer

www.environmentalvoices.org

farmwars.info — Barbara Peterson

www.flightradar24.com

www.geoengineeringwatch.org — Dane Wigington

rense.com/politics6/chemdatapage.html — Jeff Rense

newyorkskywatch.com

www.skyderalert.com

www.stopsprayingcalifornia.com

www.toxicsky.org

www.weatherwars.info — Scott Stevens

willthomasonline.net — Will Thomas

Monitoring HAARP / weather

www.aeic.alaska.edu/recent/sub/index.html — Alaska Earthquake Information Center

climateviewer.com / terraforminginc.com/climateviewer

droughtmonitor.unl.edu — U.S. Drought Monitor

earthquake.usgs.gov/earthquakes/map

earthquake-report.com

quakes.globalincidentmap.com

www.foreca.com — private Finnish weather forecasting company

ge.ssec.wisc.edu/modis-today — Modis Today

HaarpStatus.com

HAARP & HIPAS history at www.agriculturedefensecoalition.org/content/categories?q=haarp-and-hipas

ozonewatch.gsfc.nasa.gov — Ozone Hole Watch

www.patentmaps.com — Patents and inventors

rezn8d.com — The Radiation Database

radiationnetwork.com — U.S. radiation levels

Solar Cycle progression — www.swpc.noaa.gov/SolarCycle

Space and Plasma Physics spp.astro.umd.edu/SpaceWebProj/education.htm

The United Knowledge — www.youtube.com/user/TheUnitedKnowledge

www.theweatherchaser.com (Australia)

www.theweatherspace.com/haarpstatus-north-america

Health

www.cdc.gov/flu/weekly

www.electricalpollution.com

www.etcgroup.org

experimentalvaccines.org

www.organicconsumers.org/monsanto/index.cfm — Millions Against Monsanto

EM / DEW

CTBusters.com — orgone protectors

Association Against the Abuse of Psychophysical Weapons — psychophysical-torture.de.tl/Home.htm

www.biogeometry.ca — using geometry energy principles of balance biological energy systems

www.bugsweeps.com/info/electronic_harassment.html

www.cablemap.info — interactive map of fiber optics cable

"Electromagnetic Frequency Research" with Dr. Terry Robertson of the Freedom From Covert Harassment and Stalking (FFCHS); an audio of a presentation at Sonoma State University on October 2, 2010 from "Guns & Butter," KPFA 94.1 FM, Berkeley CA. ce399.typepad.com/Misc/20101201-Wed1300.mp3

www.energpolarit.com — EMF protection diodes

www.green8usa.com — bio-field protector

www.icaact.org — International Center Against Abuses of Covert Technologies (ICAACT)

www.icomw.org — International Committee on Offensive Microwave Weapons

www.lessemf.com/gauss.html — Gauss meters

www.magdahavas.com — electromagnetic effects on health

regainyourbrain.blogspot.com — On remote "non-lethal" mind control

Smart Metering Projects Map — goo.gl/mxypF

www.smartmetersmurder.com

www.submarinecablemap.com — Interactive map

www.randomcollection.info/rcp.htm — Archive of organized stalking / electronic harassment files

More Persons of Interest

Tim Ball, University of Winnipeg drtimball.com

Tom Bearden, www.cheniere.org

Nick Begich, www.earthpulse.com

David Bellamy, botanist www.davidbellamy.co.uk

Clifford E. Carnicom, www.carnicominstitute.org

Bruce Cathie, www.worldgrid.net

Wayne Hall, www.enouranois.gr/english/indexenglish.htm

Harold Lewis, UCSB emeritus professor of physics[1]

Leuren Moret, expert witness on nuclear issues

Dr. Ilya Sandra Perlingieri, author of *The Uterine Crisis* (2003)

Louis Slesin, editor of *Microwave News* (YouTube "Electromagnetic Radiation with Louis Slesin ECU #519")

Dr. Hildegarde Staninger, www.staningerreport.com

Claudia von Werlhof, www.pbme-online.org

---

1 James Delingpole, "US physics professor: 'Global warming is the greatest and most successful pseudoscientific fraud I have seen in my long life'." *The Telegraph*, October 9, 2010; Peter Gwynne, "APS responds to climate-change accusations." Physicsworld.com, October 14, 2010.

SUGGESTED INTERNET SEARCH TERMS

Scott Noble
Dr. Rosalie Bertell
Flightaware
Solar Radiation Management
Lateline
Metronome Resonance
Holes in Heaven
Norway Spiral
WeatherWar101
Project Cloverleaf
WeatherNation Climate Vlog
Skywatcher
Dave Dahl
Nanoscience
Nano-fog
KSLA News
Hologram Experiment
Metabunk
Patrick Pasin
RTL Croatia Chemtrails Report
Chemtrail Symposium
Nicola Aleksic
Pernilla Hagberg
CSEInitiative
Moscow Mountain Ridge, Idaho
Suspicious Observers
Orion Talk Radio Network
Tar Sands Pipeline
Deep Water Horizon
New Madrid Faultline
The United Knowledge
Arctic Methane Emergency Group
Enouranois
RevMichelleHopkins
TruthTV World News
Harold Saive
Oklahoma Tornado
Nexrad Weather Control
Ray Kurzweil
OpDocs:The Program
Quantum Journey
Distributed Antenna Systems
Magda Havas
Tanker Enemy
Trans Pacific Partnership
Douglas Bickford
Dutchsinse
Eric Ladizinsky
Journeyman.tv
Echelon
Daniel Hirsch
LRAD
Silent Superbug

All appendices can be found at feralhouse.com/chemtrails-appendices/